CLIMATE CROSSROADS

University Press of Florida

Florida A&M University, Tallahassee / Florida Atlantic University, Boca Raton
Florida Gulf Coast University, Ft. Myers / Florida International University, Miami
Florida State University, Tallahassee / New College of Florida, Sarasota
University of Central Florida, Orlando / University of Florida, Gainesville
University of North Florida, Jacksonville / University of South Florida, Tampa
University of West Florida, Pensacola

John M. Dunn

Climate Crossroads

Action & Reckoning in the American South

University Press of Florida

Gainesville/Tallahassee/Tampa/Boca Raton

Pensacola/Orlando/Miami/Jacksonville/Ft. Myers/Sarasota

Cover: Panoramic view from the top of Stone Mountain Park in Atlanta, Georgia. Photo by Faina Gurevich / Adobe Stock.
Book and cover design by Mindy Basinger Hill

Published in the United States of America

31 30 29 28 27 26 6 5 4 3 2 1

LIBRARY OF CONGRESS CATALOGING-IN-PUBLICATION DATA
Names: Dunn, John M., 1949– author
Title: Climate crossroads : action and reckoning in the American South / John M. Dunn.
Description: Gainesville : University Press of Florida, 2026. | Includes bibliographical references and index.
Identifiers: LCCN 2026013442 (print) | LCCN 2026013443 (ebook) | ISBN 9780813081694 paperback | ISBN 9780813075570 ebook
Subjects: LCSH: Climatic changes—Southern States | Environmental policy—Southern States | Environmental justice—Southern States | Climate change mitigation—Southern States | Southern States—Environmental conditions
Classification: LCC GE235.S68 D86 2026 (print) | LCC GE235.S68 (ebook) | DDC 363.700975—dc23/eng/20260401
LC record available at https://lccn.loc.gov/2026013442
LC ebook record available at https://lccn.loc.gov/2026013443

The University Press of Florida is the scholarly publishing agency for the State University System of Florida, comprising Florida A&M University, Florida Atlantic University, Florida Gulf Coast University, Florida International University, Florida State University, New College of Florida, University of Central Florida, University of Florida, University of North Florida, University of South Florida, and University of West Florida.

University Press of Florida
PO Box 140239
Gainesville, FL 32614
floridapress.org

GPSR EU Authorized Representative: Mare Nostrum Group B.V., Doelen 72, 4831 GR Breda, The Netherlands, gpsr@mare-nostrum.co.uk

CONTENTS

CLIMATE CROSSROADS

INTRODUCTION

The first attack on Fort Sumter occurred on April 12, 1861, as troops of the Confederate States bombarded the two-acre island fortress in South Carolina's Charleston Harbor, launching a civil war against the United States of America.

Since then, another kind of slow-moving assault has been battering the fort day by day that may prove more destructive than anything Confederate soldiers could have unleashed. Instead of iron cannon balls, rising seawater and ever more potent stormy seas and wind have been eroding the shell-pocked, five-foot-thick brick walls, at times forcing the US Park Service to close the cherished national monument. There are no signs of a ceasefire.

What is happening there in Charleston Harbor, of course, is hardly unique. Along the entire coastline of the American South numerous other historic sites, homes, military bases, cemeteries, commercial and government buildings, essential waterfront infrastructures along with a vast myriad of natural landscapes—wetlands, marshes, estuaries—are all in peril of being swamped, salinized, washed or blown away by extreme weather events. Scientists have a word for what is causing these turbulent events—*forcings*, that is, influences, human-caused or natural, powerful enough to change the climate by altering the balance between the amount of solar energy entering the atmosphere and energy that is leaving. And the grandest of these by far is the human-caused spewing of excessive greenhouse gases (GHGS) into the atmosphere where they, along with other vapors, capture heat radiating from the earth and thus slowly roast the planet. For the moment, this greenhouse explanation for the phenomenon known as climate change is the best there is, according to a consensus reached by an estimated 99 percent of climate scientists around the planet who consolidate and analyze their findings for the International Panel on Climate Change (IPCC).[1]

Even if, as some researchers suggest, that level of agreement may be overstated by a few percentage points, it is nonetheless true that almost all climatologists keep confirming that disruptive and dangerous climate changes are occurring, primarily from human actions and choices. This isn't news, however. Signs of climate disruption have been making themselves apparent for more than a century—yet too few people noticed or took them seriously. For some Americans, the idea that something as huge and complex as climate could be changing for any reason remains unbelievable. All the same, massive scientific evidence contradicts such disbelief. Moreover, numerous scientific assessments indicate the American South may be the nation's region most vulnerable to climate change and that the hardest hits are yet to come. It's not just a matter of sea level rise and extreme weather events, which are happening around the world. The US South, say social science researchers, also has a disproportionate share of social, economic, and infrastructure vulnerabilities to the impact of climate change when compared elsewhere in the nation. Too many Southerners live below the poverty line, reside in substandard housing, and lack adequate child and maternal care. Not enough of them have health and property insurance. Among the most vulnerable are those Southerners impacted by the legacies of enslavement and racism who also live in polluted communities. So far, conclude researchers, preparations taken by southern state governments for disaster and recovery as the climate crisis unfolds have been lackluster. Such neglect likely means higher costs in the future.

But where is the South? Some may respond it's the 11 states that composed the former Confederacy—Louisiana, Mississippi, Alabama, Georgia, South Carolina, North Carolina, Virginia, Tennessee, Texas, Florida, and Arkansas. Social scholars who study the South as a distinct cultural force in America aren't so sure. They wonder, for instance, if Oklahoma, Delaware, Maryland, Missouri, and West Virginia—states that once enslaved people but remained in the Union and culturally resemble their neighbors in nearby states—should be considered Southern. On a map, Florida unques-

On August 20, 2025, Category 5 Major Hurricane Erin located off the Atlantic coast of Florida heads northeast. iStock.com/FrankRamspott.

tionably lies in the South, and it was also the third state to secede from the Union, but today many Southern purists say the Sunshine State lost its bona fides long ago. After all, most Florida residents today weren't even born in the state. *Travel South,* which calls itself "The Official Regional Destination Marketing Organization of the Southern USA" omits Florida—a former member of the Confederacy—from its scope of interest. The US Census Bureau and many social geographers, however, do consider Florida part of the South precisely because of its history and regional location and add a few border states to their mix. It is this more comprehensive perspective, perhaps, that best fits any examination of the entire region as well, as the South confronts climate change.

A Troubled Legacy of Agricultural Greatness and Shameful Exploitation

One way to understand the interplay of what it means to live in the South is by taking a historical look through a lens of change and resistance to change. Arguably, major social and environmental shocks arrived with white colonists, whose settlement practices involved dispossessing, removing, enslaving, and killing by bullets and germs Powhatan, Nottoway, Meherrin, Siouan, Muskogean, Calusa, Tequesta, Creeks, Choctaw, Chickasaw, Seminole, Creek, and other Indigenous peoples of the South. The southern latitudes of North America offered long growing seasons and fertile soils, which made the region ideal for large-scale plantations and a profitable cultivation of tobacco, cotton, rice, and sugar cane. In time, the labor of enslaved Africans and their descendants made the South the economic engine of the entire nation during the first half of the nineteenth century. If the South were a separate nation in that period, it would have been the fourth-richest region in the world, outdoing all European countries except England.

The old Southern hierarchical plantation structure of top-down authority fits compatibly with the corporate management structure of many modern corporations. Contemporary research also suggests that most plantation owners used sophisticated business practices, such as calculating depreciation of their enslaved workers. "They appraised their inventory at market value, compared that with its past market value to assess appreciation or depreciation, calculated an allowance for interest, and used this to determine their capital costs," says Caitlin Rosenthal, a history professor at the University of California, Berkeley. "It's really as sophisticated as what most firms do today."[2]

The whole point of using these advanced procedures was optimization of resources—a forerunner of routine efficiency practices found in many modern business settings that analysts tend to applaud. In the Old South, optimization meant using human beings as capital and market goods.

That large-scale labor exploitation also damaged the environment. "On its leading edge, slavery laid waste to fragile ecosystems," writes David Silkenat, author of *Scars on the Land: An Environmental History of Slavery.* "On its trailing edge, slavery left eroded hillsides, rivers clogged with sterile soil, and the extinction of native species."[3]

The Civil War further exacerbated environmental destruction. Swaths of Southern forests were blown away and torn apart by cannon balls and bullets. Fires took a toll, too. Nobody knows how badly the region's rivers and soils were contaminated by the tons of lead in spent bullets and the decay of putrefying corpses.

There was, of course, a yearning for rebirth in the South following its defeat in 1865, and many of the region's leaders wanted everyone to know about it. There's now a New South growing and building, they proclaimed. Boosters of the South have been repeating this mantra ever since.

The grandfather of Southern boosterism was surely Henry Grady from Athens, Georgia, "the Voice of the South." In 1886, Grady, the featured guest speaker at a New York meeting of the New England Society, assured his audience the South had changed. "There was a South of slavery and secession—that South is dead," he reported. Moreover, "There is a South of union and freedom—that South, thank God, is living, breathing, growing every hour." And this rising phoenix, he added, promised to deliver not just the fruits of agriculture, but also a "diversified industry that meets the complex needs of this complex age."[4]

At the time of Grady's speech, however, conditions in the South didn't look too promising. Devastated physically, economically, and psychologically, Southerners found themselves in a world still chiefly reliant on agriculture. Though railroad tracks were spreading across the region as commerce struggled to make a comeback, poverty and economic hardship were widespread. Although the war's outcome meant Black Southerners were now free, they found themselves trying to survive in a traumatized society in which they were still not equal to whites and often in peril of being ensnared in one of the South's numerous convict-leasing systems or trapped in debt peonage resulting from lopsided sharecropping contracts. Either way, the South's commercial class was assured of at least a promise of an ample supply of cheap, or even once again, free labor.

GHGS on the Rise

It is improbable that any Southern booster had the vaguest inkling that the century-old industrialization process they all wished for bore a hardly known side effect that would eventually imperil the South and the rest of the planet. Simply put, the new industrial processes that were essential to the growth of manufacturing in the United States and Europe thrived on the burning of fossil fuels. This meant ever greater quantities of carbon dioxide were polluting the air, which along with naturally occurring gases such as water vapor, methane, and nitrous oxide, would become known today as greenhouse gases. Few people understood in those days the deleterious impact these gases could have on the planet. Today, however, many people can recite climate change's basic-narrative origins: Sunlight strikes the Earth. Some of that energy is reflected up into space, the rest is absorbed. Warmed-up surfaces of Earth later radiate this solar energy into the atmosphere. There it is trapped by GHGs, including those that have been accumulating rapidly since the late 1700s from human beings burning plants, wood, coal, kerosene, gas, and whale oil for warmth, cooking, protection, weaponry, and land clearance. Over time a huge invisible and ever-growing heat blanket high above has been literally warming the planet. The seas are warming, too, which means water molecules are expanding and causing a rise in sea levels. Climbing temperatures also mean that land-based glaciers and ice sheets are melting and draining offshore, also causing seawater to rise.

The First to Figure It Out

Even in the early industrial days, however, a handful of scientists were already suspecting that too much of an industrial air pollutant such as carbon dioxide might become problematic. In 1824, for instance, Joseph Fourier, a French mathematician and physicist, was the first person to suggest the Earth's atmosphere could raise or lower the planet's temperature. In the United States, Eunice Foote, an amateur scientist, shared her research at the 1856 annual meeting of the American Association for the Advancement of Science (because of sexist conventions at the time, she had to solicit a male colleague to present her paper). The audience learned about her experiments showing how rising levels of carbon dioxide and water vapor could increase atmospheric temperatures. Three years later, Irish scientist John Tyndall, using more sophisticated experimentation, was able to confirm that certain gases could block infrared radiation and change the climate. In 1896, Svante Arrhenius, a Nobel Prize–winning Swedish chemist, used the principles of chemistry to estimate how and to what extent an increase of CO_2 in the atmosphere would boost surface temperatures on earth.

The findings of these scientists interested many in some scientific circles, but beyond that almost anyone else who gave fossil fuels a second thought was more interested in exploiting them. In the South, one of the first of its own ample supplies of these potential fuels to be developed for industrial purposes was coal. At first, the business of extracting it got off to a slow start during the 1700s in the mineral-rich Appalachian Mountains that run from Pennsylvania to northern Alabama, but it later took off spectacularly in the 1870s after railroad builders at last penetrated the Cumberland Mountains, providing swift transportation for coal shipments to the nation's industrial centers and paving the way for the growth of steel production in Birmingham, Alabama.

By then demand was also mounting for another fossil fuel the Appalachians seemed to have plenty of—petroleum, or crude oil. Its existence had been known since 1818 when oil first gushed to the surface in Kentucky after two men drilling salt wells along the South Fork of the Cumberland River struck crude oil instead. West Virginia's first major oil well debuted in 1860, and six years later the gooey black stuff was also discovered in Tennessee, the same year Texas drilled its earliest, but modest producing, well in Nacogdoches County. By the end of the nineteenth century, oil wells were pumping "black gold" up from the ground in coal-rich mountain states such as Kentucky, West Virginia, and Tennessee.

In the next century, more oil extraction operations appeared in other Southern states, including Arkansas, Oklahoma, North Louisiana, and East Texas. In 1943, the Humble Oil and Refining Company sank Florida's first oil well.

Thus, the South was making arduous, yet meaningful, strides to join in a rapid technological, social, economic, and political revolution occurring across the Western world. Its role seemed ordained to be a supplier of fossil fuel, just what the new machine age needed.

Burgeoning Climate Research

Of note are significant scientific findings that kept arising during decades of fossil fuel production, such as those in 1938 of the Canadian steam engineer and inventor Guy Callendar. By compiling temperature measurements from 147 stations around the globe over a 50-year period beginning in the nineteenth century and making complex calculations, Callendar determined global land temperatures had risen during that period and suggested that an increase in CO_2 emission was responsible. Although overlooked at the time, his work has been mostly validated by modern analysis.

Climate predictions, however, were of little concern to the nation's leaders as the United States mobilized for war in 1939. The huge federal demand for weapons and other war-related goods and services revitalized the economy. Though the South, too, felt the impact of emergency-spurred industrialization, it still seemed stuck in an economic rut. Opposition from the region's powerful white patriarchs to federal efforts to provide federal aid during the ongoing economic depression to not only white people in need, but also those trapped at the lowest level of the Southern caste system, only made the situation worse.

Meanwhile, GHGS kept polluting the skies above the American South and the rest of the world, as scores of nations during World War II extracted fossil fuels to wage war. They continued during the postwar 1950s, as nations rebuilt their war-blasted economies and landscapes, and they have been increasing ever since.

A President Receives Early Warning of Climate Risk

In February 1965, President Lyndon Johnson, a native of oil-rich Texas, received a report from a group of climate scientists that presented research on the risks of rising levels of CO_2 in the earth's atmosphere caused by the burning of fossil fuels. The researchers were remarkably prescient: "Through his worldwide industrial civilization, man is unwittingly conducting a vast geophysical experiment. Within a few generations he is burning the fossil fuels that slowly accumulated in the earth over the past 500 million years. . . . The climatic changes that may be produced by the increased CO_2 content could be deleterious from the point of view of human beings. The possibilities of deliberately bringing about countervailing climatic changes therefore need to be thoroughly explored."[5]

Johnson authorized the report to be made public, and he also mentioned the danger of excessive carbon dioxide in a speech to Congress. However, at the time, his administration was more focused on solving pollution problems such as dirty air and water. Thus, nothing was done about climate change. Nonetheless, Johnson made history as the first American president to be officially warned about a slow-moving danger to national security and to publicly acknowledge its existence—60 years ago.

Two Societies

A little more than three years later, on April 4, 1968, Americans and people around the world reeled with shock after a white shooter assassinated Martin Luther King Jr., the most well-known public face of the civil rights movement and apostle of nonviolence, in Memphis, Tennessee. King's murder occurred 35 days after the release of a report conducted by a special 11-member National Advisory Commission on Civil Disorder established by Johnson to examine the causes of racial discord in 23 American cities, such as Detroit, Newark, and Cincinnati. The nation, warned the commission, was "moving toward two societies, one black, one white—separate and unequal."[6] And it was white racism "accumulating in our cities since the end of World War II" that was responsible for the riots. Among the "ingredients of this mixture," said the commission, "were ongoing discrimination and segregation in education and housing, black in-migration and white exodus from major American cities."[7]

Accumulating wealth was another hurdle for most Black Americans. Many were descendants of the enslaved who were mostly unpaid for their labor, and thus unable to pass on wealth to heirs. In the first half of the twentieth century, Jim Crow economies kept millions of African Americans poor in the South. Making matters worse, during the second half of the twentieth century, most Black families were unable to take advantage of the best way to accumulate wealth in America—home ownership. Unlike their white counterparts, they were largely left out of the housing market because of redlining, a practice used by the federal officials and private lenders who literally drew a red line on a map to designate neighborhoods—usually Black ones—as too risky for approving a loan for home purchases. This meant a great many Black Americans were doomed to pay onerous rents all their lives in segregated neighborhoods, which often were plagued with crime, dangerous highway traffic, limited government services, and polluted air and water. Without mortgages, they were stuck in place. Redlining, then, was a policy that directed millions of Black Americans and others into hazard zones where they spent their days near stinking oil refineries, leaky landfills, busy interstates, aging nuclear power plants, and chemical production centers, increasing the likelihood that they and their offspring would live shorter and sicklier lives than most white Americans. Sometime, these toxic large-scale enterprises appeared in thriving, relatively unpolluted Black communities at the request of outside rich and powerful interests that viewed local inhabitants as unimportant or expendable. The effects of climate change, if not mitigated, are expected to worsen these iniquitous conditions.

Thanks to new federal laws and changing economic conditions, however, life for African Americans slowly improved during the decades following the heyday of the civil rights movement. "It is undeniable that the economic fortunes of African Americans changed dramatically during the twentieth century," concludes Thomas Maloney, an economics professor at the University of Utah. "African Americans moved from tremendous concentration in Southern agriculture to much greater diversity in residence and occupation. Over the period in which income can be measured, there are large increases in Black incomes in both relative and absolute terms. Schooling differentials between blacks and whites fell sharply, as well."[8]

However, cautions Maloney, "this century of progress ended with considerable gaps remaining between African Americans and white Americans in terms of income, unemployment, wealth, and life expectancy."[9]

Southern Growth

Despite its problems, the American South in recent years has seemed to millions of Americans of all races and millions more living abroad an attractive and desirable region to live, work, and invest. By 2019, in fact, according to the US Census Bureau, the South led the nation in net population gain from other parts of the nation, with Florida receiving the biggest number of new residents. Why wouldn't people move here? asked legions of smiling community leaders at countless Southern chambers of commerce and economic development councils. Like Henry Grady, they, too, wanted to feature their communities in a favorable manner with their industry recruitment slide presentations.

In recent decades, manufacturers and businesses from the North and abroad have relocated to the South, drawn by a constellation of enticing factors. Land is relatively cheaper. Energy costs are lower, too. Environmental regulation and enforcement are lax. Major interstates, airports, and seaports are never far from the South's urban centers. The avowed corporate climate is decidedly "business friendly." State and local governments offer an array of incentives to relocate businesses and industries, such as taxpayer-financed industrial parks, access roads, worker-training subsidies, reduced utility rates, and in some places cheap or even zero taxes. The South is also a region of "right-to-work" states.

And so, millions, responding to the lures, have been moving to the South where they live on landscapes scraped of natural vegetation in oversized, look-alike, energy-intensive houses supplied by electricity derived mostly from fossil fuels. The roads are congested with increasing numbers of CO_2-releasing cars because comprehensive mass transit hardly exists. Pro-growth voices want even greater populations to find their places in the sun.

Amid all this development have also come nonstop, and increasingly alarming, warnings from the IPCC and other scientific bodies that the climate of the world is changing. The evidence has been hard to miss. Year after year, New Orleans, Charleston, Miami, and innumerable other cities and towns on the coasts of the Gulf of Mexico and the Atlantic Ocean have slowly been swamped. Tropical storms keep growing more powerful.

Hot temperatures are setting new records. All the world has been feeling the impact, of course, but in the United States, experts kept saying those populated regions in the American South—because of its geography and its enduring economic, racial, and social hierarchies—were at the greatest risk.

Southern slang has an expression for this kind of coming train wreck: *Hot Mess*. And the one that's coming to the Southland is likely to be of epic proportion.

Along with people everywhere during the age of climate change, Southerners now find themselves at a crossroads—a netherworld seeped in the folklore of the region. One of the South's most enduring myths is the tale of Robert Johnson, a young African American Mississippi bluesman who met the devil one night in the 1920s and traded his soul for fame. In 1938, however, his good fortune ended abruptly when he died at age 27, just as plans were underway for him to perform at Carnegie Hall. Johnson's legacy, though, lives on as a reminder of what happens when we head the wrong way at the crossroads.

"Throughout the Twentieth Century, the South stood at the crossroads," argued American historian and museum curator Pete Daniel in 1996. Would the region, he wondered: Continue segregation? Abandon the sharecropping system and become mechanized? Rid itself of violence, lynching, and intimidation? Keep mythologizing its past or move on and create a better society? Remain "a creative backwater" or produce new art? Lose its unique accent or cling to it? And finally, "Continue as the South or become a replica of the North?"[10]

Today, the South is neither. Rather, it is a blend of both worlds, a region where winds of change are both resculpting the landscape and stoking long smoldering fears and grievances. Today's South sees a renewed civil rights movement whose eye is on new prizes—this time a broad array of social justice goals, including reckonings for those unduly harmed by environmental abuse expected to worsen in the era of climate change.

Traveling alongside, meanwhile, is another robust movement filled with climate change activists and their allies in local governments scattered across the South. Though the goals of both camps aren't always the same, they face the same opposition—a juggernaut of powerful corporate and industrial forces, ideologues, religious and political conservatives and reactionaries, fossil fuel industry lobbyists, state legislators, and governors all determined to block, if not destroy, the major goals of the activists, especially those meant to remedy the causes and impacts of climate change in all its forms.

Despite all the sound and fury by their adversaries, climatologists are quick to point out, "And Yet the Sea Rises." And, they add, humans are to blame.

Peering at how the drama is playing out in the South is the primary focus of this book. Its author, a native Southerner who has lived through many of the great modern upheavals and transformations in the American South, hopes to demonstrate that what this region does, or fails to do, about climate change matters greatly. As the great African American scholar W.E.B. Du Bois reportedly observed during the tragic days of Jim Crow, "As the South goes, so goes the nation."[11]

Hurricane Debby flooded homes and cars in the Laurel Meadows community in Sarasota, Florida, on August 6, 2024. iStock.com/Bilanol.

ONE
HOT TIME IN THE SOUTHLAND

In 2017 New South boosters could hardly have been pleased with the big takeaway coming from the Climate Impact Lab, a consortium of economists and policy experts from the University of Chicago, the University of California, Rutgers University, and the Rhodium Group, an independent research firm. These researchers had used computer-modeling which for the first time allowed climate economists to develop a more nuanced understanding of climate change's impact at the local and regional levels. Focused on six factors—temperature, rainfall, the effect of carbon dioxide on agriculture, mortality, crime, labor, and energy demand—their study shows that the South, particularly fast-growing Florida where most residents live along the coasts, can expect to be hit hard by weather extremes.

As journalist Robinson Meyer, writing in the *Atlantic*, puts it, "The American South will bear the worst of climate change's costs . . . [A]cross the country's southern half—and especially in states that border the Gulf of Mexico—climate change could impose the equivalent of a 20-percent tax on county-level income, according to the study. Harvests could dwindle, summer energy costs soar, valuables will erase real-estate holdings, and heatwaves will set off epidemics of cardiac and pulmonary disease."[1]

The editors of *Popular Science* warned, "If you live in the South, climate change could kill your economy."[2] It could quite possibly kill a lot of people, too. According to a study published in *Science* on June 30, 2017, mortality rates, especially those caused by heat, could rise by as many as 80 deaths per 100,000 in every county between El Paso, Texas, and Charlotte, North Carolina.[3]

The South's geography, especially its low-lying coastal areas, obviously makes it extremely vulnerable to climate-related hazards: hurricane winds, drought, flooding, and sea level rise. Making matters worse, the region also suffers from a high degree of social vulnerability for those disadvantaged by race, ethnicity, poverty, and age who live in environmental hazard zones. For them, climate change is likely to exacerbate the damage and misery already in play. The most vulnerable of these kinds of places found in the South, according to the Environmental Defense Fund and Texas A&M University, are industrialized communities situated along the Gulf Coast, ranging from Corpus Christi, Texas, to Mobile, Alabama—a corridor tormented by flooding, hurricanes, and rising seas.[4]

Wetland Worries

Signs of distress from a warming planet are now easy to see, even if you're not a scientist. Along the Louisiana coast, for instance, a patch of land equal to one football field is swamped every one hundred minutes. The threats Louisiana is facing "seem like a list of plot lines from a dystopian disaster movie," concluded Mark Schleifstein of the New Orleans newspaper the *Times-Picayune,* after he read alerts from the IPCC in March 2022.[5] Ghost forests—strands of swamped, skeletal-looking dead cypress and other trees and browning wetlands—are growing, the victims of salt-poisoning. You can see them not only in Louisiana and other coastal states, but also along the Suwannee River in Florida, or at Georgia's Jekyll Island, or in the marshy, low-lying coastlines of the Carolinas and Virginia.

Wetland specialists have long maintained that the lower 48 states may have lost more than 50 percent of their coastal wetlands since the eighteenth century—though some recent reports suggest this loss may be closer to 21 percent. All the same, tens of thousands of acres are still being wiped out every year thanks to drainage, fill operations, pollution, agriculture, development, and the impact of climate change. The significance of that loss looms large in the Southeast, home to 81 percent of coastal wetlands in the contiguous United States.[6]

Wetlands command our attention because they serve as guardians for a big swath of the natural world, ranking among nature's most biologically diverse ecosystems. They not only provide habitat for animal and plant life but also serve as buffers that protect the shore from increasingly destructive storms. They function, too, as reservoirs, absorbing and retaining huge amounts of freshwater. In the South, wetlands serve as living icons of an outdoor way of life that provides stunning natural beauty and the bounty of delectable seafood. Wetlands adorn the coastlines of Florida, Mississippi, and Louisiana and range from the Chesapeake Bay to North and South Carolinas Low Country. The South wouldn't be the same without them.

Another benefit of wetlands is their diverse array of plants that capture massive amounts of carbon dioxide. For several years, scientists have studied whether these carbon catchers would also spur new plant growth, thus expanding wetlands and thereby moderate sea surges. But, sadly, a study sponsored by the Smithsonian Environmental Research Center released in 2022 indicates there may be a limit to how much wetland plants can grow, especially since seawater keeps swamping wetland plants, making it harder for them to thrive, or even function.[7]

A New Era of Killer Storms

As wetlands shrink and their utility as storm barriers diminish, many hurricanes are becoming more powerful and destructive. Under a certain set of circumstances, we may not necessarily see more tropical storms, but rather a greater number of deadly Category 4 and 5 tempests. Some scientists even say there is now a need for a Category 6 rating for monster storms with wind speeds above 192 miles per hour. When tropical storms move across ever-warming oceans, they absorb greater amounts of heat and water vapor, fueling their potency. "It's not sea level rise that's the biggest existential threat," emphasizes L. Donelson Wright, the former Director of Marine Science at William & Mary's Virginia Institute of Marine Science. "It is the combination of storm surge and torrential rain that now cause the most devastating coastal events associated with climate change."[8]

And this deadly combination poses heightened danger especially for communities of Florida's Gulf Coast, which unlike the state's eastern Atlantic side, sits upon a wide, shallow continental shelf that functions as a huge underwater runway that amplifies the ferocious kinetic power driving a sea surge spawned by a hurricane. "No longer should anyone on Florida's Gulf Coast think that just because you rode out the last storm, it is safe to do so again," warns Wright. "Don't plan to ride out a new hurricane; they've become too dangerous. The landward penetration of the storm surge with a Category 3 storm is five miles. Don't mess with it. Heed the warnings. Don't wait another day to make your decision. You don't have 'another day' anymore."[9]

Wright's conviction deepened in October 2022, when Hurricane Ian took 149 lives in the Southwest Florida coastal cities of Fort Myers, Cape Coral, and Naples. "If the storm hadn't made a right turn when hitting the western edge of Cuba before heading north to Florida," he says, "it may have struck Tampa Bay instead where a 23-foot surge was expected on the downtown streets of Tampa." Another anticipated problem was that Tampa's streets would be swamped with "dirty water with alligators, snakes, and deadly water-borne pathogens."[10] This would have been a shattering blow to a community already plagued by sewage spills and freshwater supply woes.

Threat from Below

Water contamination problems, meanwhile, can be expected to persist if not grow worse as encroaching seawater, tides, and storms continue contaminating freshwater sources along the coastal states. Seawater has been creeping north into the Mississippi River for years. In many places in almost pancake-flat Florida, saltwater is salinizing soil as it infiltrates freshwater water pipes and overwhelms aging sewer systems, causing them to back up and contaminate freshwater supplies.

Seawater is also moving upward from absolute darkness, literally deep far beneath the feet of Floridians, ascending through karst, that is, the Swiss cheese–looking porous layers of underground limestone reservoirs known as aquifers that provide 90 percent of Florida's drinking water. All the while, this unimaginable uplifting mass of seawater is exerting skyward pressure against vast overhead layers of brackish water and quadrillions of gallons of fresh groundwater.

Making the situation worse, a growing number of Floridians keep pumping freshwater from that same aquifer in ever-increasing quantities for agriculture, drinking water, hydration of grassy lawns, and many commercial purposes. Removing freshwater this way only makes it easier for rising seawater to reach the surface.

As many South Floridians know, the ever-surfacing freshwater ground table in their neighborhoods is infiltrating the drain fields of leaking septic tanks and aging sewer lines, thus becoming contaminated and unfit to drink. Small wonder that increasing numbers of coastal Florida cities such as Hallandale and Fort Lauderdale have been searching further inland to sink new and safer municipal wells.

The sea level is also rising faster in South Florida than expected, according to recent data from the National Oceanic and Atmospheric Administration (NOAA). In fact, the federal agency reports that over the past two decades it has risen 6.5 mm to 9.4 mm each year, far faster than the 3.9 mm per year historical rate. This means trouble for some mangrove forests—tropical and subtropical groups of trees, tangled roots, and shrubs in the coastal intertidal zone—along the Florida coast. Unless mitigation efforts, such as huge infusions of freshwater, are undertaken by engineers, say NOAA researchers, rising seas will outpace the ability of mangroves to accumulate enough sediment to sink their roots and thrive. Any loss of coastal mangrove forest, which also has been inching northward because of rising temperatures, would weaken nature's barriers to windstorms and surging waves. Fewer mangroves also suggest less intake of carbon dioxide.[11]

Rearranging the Natural Order

A warming world is altering the geographic distribution of plants, animals, and other life forms big and small. Studies reveal much of the South is already undergoing this process in multiple ways, and one of the most noticeable is what scientists call tropicalization—the transformation of temperate ecosystems by northern-moving tropical organisms being spurred into action by warming temperatures.

The normal transitional areas between tropical and temperate zones are usually in a state of flux owing to variations such as those caused by changes in rainfall and water chemistry along with the relentless disruption by human activities. Until recently these vacillations were not too dramatic because cold temperature extremes occurred often enough within the transition zones to kill off encroaching invasive tropical life forms and keep them from colonizing the temperate world. Today, however, such extremes are becoming fewer, if not disappearing, thus in effect opening the borders to migrating tropical and subtropical life forms that may populate new territories and even replace and re-create the ecoscape they are entering.

Signs of tropicalization are everywhere. In Florida, spoonbills, wood storks, Glossy Ibis and many other wading birds, manatees, sea turtles, alligators are spreading their domain northward; so are cold-blooded American crocodiles and alligators, along with the omnivorous pythons, those exotic predators that have ravaged wildlife in the Everglades, swallowing even deer. Tiny life forms have joined migration too, including mosquitoes that carry diseases that spread to people, such as West Nile virus, St. Louis encephalitis, and eastern equine encephalitis. By June of 2023, health officials had detected cases of malaria spread by mosquitoes along the Gulf Coast of Tampa and coastal Cameron County, Texas. The outbreaks marked the first time in 20 years of such a spread. Other diseases common to warm climates, namely, dengue fever and *Vibrio vulnificus*, the so-called flesh-eating bacteria that thrives in warm, brackish floodwaters, could also become more common in coming years.

Loss of habitat along with inability of wildlife and flora to quickly adapt to a changing climate are other potentially dark outcomes. Coral reefs across the Caribbean and the Florida Keys, for instance, are dying from a host of malign influences, chief among them marine heat waves with abnormally high temperatures.

These warming waters, along with nutrient run-offs, are boosting the production of other life forms, such as thickening Sargassum seaweed and other macroalgae which foul South Florida beaches, costing local communities billions of dollars in economic loss when tourists stop coming.

Creatures that don't or can't adapt or migrate, of course, are in trouble. The feral horses on the Shackleford Banks of North Carolina, for instance, are now having trouble finding potable water and may have to be relocated. Along the coasts of South Carolina and elsewhere, rising seas are washing away nesting habitat for endangered loggerhead sea turtles. In the Florida Keys, the diminutive key deer keep watching the rising sea nibble away at the only place they've ever lived.

Hybridization is also happening. In 2025, research at the University of Texas confirmed through genetic analysis that a strange bird spotted by a birder in San Antonia was the hybrid offspring of a male Blue Jay and female Green Jay. Their union may be the result of their respective ranges—until now separated over millions of years—merging in recent years because of warming temperatures. The new "grue-jay" may be the first, or one of the first identified, cases of its kind related to climate change.

Southern landscapes are changing, too, thanks to ongoing alterations to regional soil composition, evaporation, and precipitation patterns, along with temperature changes, droughts, fires, floods, erosion, wind, and salty water. It may be a planet-heating agent, but Co2 is quite possibly leaving a different kind of impact on flora. All that trapped carbon dioxide in the atmosphere, as some agricultural scientists have maintained, is expected to boost photosynthesis, which helps plants grow quicker, bigger, and more effectively. Some experts think by 2050 many forests could be growing faster than ever—maybe 18 percent swifter—thanks to the extra gas. According to one study, the speediest plant in the woods could be poison ivy, expanding 80 percent faster and producing more potent amounts of urushiol, the oily substance that causes skin rash, bumps, and itchy blisters.

On the other hand, a rise in heat may make it harder for some plants to get all the carbon dioxide they need. Excessive heat, say Australian researchers, can curtail the efficiency of the enzymes responsible for photosynthesis and thus make it harder for a plant to control water loss and regulate its uptake of CO_2. Call it plant stress.[12]

The USDA, working with University of Washington researchers and other scientists, has concluded that a rise in the amount of gas emissions can reduce nutrient values in rice, which is a troubling forecast for rice producers in the Mississippi River Delta area, the fourth-biggest exporter of the grain in the world. Ever-more-powerful storms in the South play havoc with crops, too. In 2018, for example, Georgia received a glimpse of possible hazards yet to come when farmers suffered $2.5 billion in crop losses thanks to the destructive force of Hurricane Michael. Florida citrus farmers have similar woes.

Hotter temperatures, of course, hurt livestock, too. Not only do cows and poultry suffer, but heat-related stress can weaken their bodies, rendering them less productive.

The Cost of Heat on Humans

In summer, humid and hot conditions in the South always make life uncomfortable, if not unbearable, for millions of sweaty people. Air conditioning was one of the technologies foundational to Florida's post-WWII population boom, but you can't take it outside. The threat of heat exposure in the South is now so high that it has caught the attention of not only the Occupational Safety and Health Administration and a spate of southern regulatory agencies, but also many commercial institutions. And once again, all eyes are on the Sunshine State.

"No state in the US has a more dire climate change outlook than Florida, a state that ranks as a top-five worst state for both extreme heat and flooding," according to a report from *Policygenius,* the online insurance broker. "Even in a low-emissions scenario, Florida is still expected to see a whopping 86 days with the heat index above 100 degrees by 2050, a 61-day increase over the historical average."

Analysts at the insurance firm also point out that Florida has underlying structural problems that could make outcomes even worse in coming years. For one thing, the state has the "6th most vulnerable heat-population," a big percentage of whom are seniors.

Florida isn't the only southern state on the road to hellish days. In second place on the *Policygenius* list is Mississippi, followed by Louisiana, Texas, and Arkansas.[13]

What may shock Southerners—and those wishing to move to the South—is the blunt warning from the Union of Concerned Scientists that if we take no major corrective action, the South, by the end of the century, will not only be the hottest region of North America, but also a region where people suffer temperatures with a heat index above 100 degrees anywhere from 100 to 200 days a year. According to Climate Central, another scientist-packed research group, severe living conditions are in store for the Southeast and the Gulf Coast, where the combination of heat and humidity is projected to cross thresholds dangerous for human health within the next decade.

For the moment, Texas ranks number one in the nation for coping with summer drought. By 2050, the state, along with several others, is likely to be dangerously threatened by wildfire. Studies show a similar fate awaits Florida, North Carolina, Georgia, Arkansas, Alabama, Louisiana, Arkansas, and Mississippi, where an estimated 35 million people live in locations with an overlap of fire-prone wildlands and developed residential areas.

Too Much Rain and Not Enough

The many ways rising heat will impact weather systems are harder to predict. Along with other factors, for instance, it is expected at varying times and places to accelerate evaporation that could contribute to drought and parched landscapes. Paradoxically, however, it is likely there will be times when enhanced evaporation will increase the chances of bouts of heavy and often sudden and destructive rainfall. This phenomenon may explain the recent rash of destructive floods in many inland southern areas, such as the Appalachian communities. In the summer of 2022, Kentucky, for example, experienced the worst flooding in its history, quite possibly the byproduct of hotter temperatures. Roadway flooding could be so bad in the next two decades that many southern roads will be impassable, according to the October 2021 *3rd National Risk Assessment: Infrastructure on the Brink*, produced by the nonprofit First Street Foundation. "These areas tend to be in regions with well-established flood risk, such as coastal flood plains along the Gulf and Southeastern coasts of the US, but also in less well-known flood zones, such as in the Appalachian Mountain regions of West Virginia and Kentucky." To that point, 17 of the top 20 counties in the US which are most at risk for flooding are in Louisiana, Florida, West Virginia, and Kentucky. Additionally, the top cities at risk of flooding persistently show up in the states of Louisiana, Florida, Texas, and South Carolina. The analysis further uncovered a high degree of vulnerability in some of the major population centers in the United States, including New Orleans, Miami, Tampa, Charleston, Chicago, and Los Angeles.[14]

As conditions grow grimmer in the South, residents of some northern states may benefit from rising temperatures, at least for a while, say the researchers at Climate Impact Lab. Growing seasons could lengthen in those states, and milder temperatures encourage economic activity and fewer workdays lost to snowstorms. Some cities such as Buffalo, New York; Duluth, Minnesota; Madison, Wisconsin; Detroit, Michigan; and Burlington, Vermont, are already advertising themselves as "climate havens" for those fleeing the South and Southwest.

Misery Along the Mississippi

Climate change is adding extra burdens to residents in southern locales already in an environmental decline. This is glaringly evident in Louisiana, especially along the Mississippi River Delta, a region already facing a slew of destructive forces whose origins are both natural and anthropogenic. For years, scientists, coastal residents, environmentalists, elected officials, Indigenous people, commercial fishers, and engineers have been struggling to repair and protect the state's beleaguered coastline. Much of their time and resources is spent on the Barataria Basin—a vast area consisting of over 1.5 million acres of lakes, marshes, wetlands, and estuaries, located both south and west of New Orleans. Bordered by the Mississippi River on the north and south, the basin has been in ecological decline for decades.

And now, owing to climate change, sea level rise, higher temperatures, saltwater intrusion, and more intense hurricanes are adding to the toll. Vanishing wetlands, the product of both natural and human forces, also contribute to the problem of subsidence—the sinking of earth, a process which allows deeper flooding from tides and sea level rise that disrupt hydrology and destroy plant life. All this leads to even more wetland loss, thus, a self-enforcing loop of destruction.

Human engineering helped set the stage for this decline. Begun over three hundred years ago primarily to control flooding, the river's extensive levee network confined the mighty river and turned it into a major canalized transportation waterway that empties into the Gulf of Mexico. By keeping the Mississippi in a straitjacket, the engineered drainage system has controlled much flooding in populated areas, but it has also prevented the river from depositing nutrient-rich sediment into the basin, as a natural flowing river would. Without periodic replenishment, both the basin's wetlands fertility and its land surfaces have declined over the decades, reducing both habitat and sea-surge protection. Levees also altered the duration, timing, and frequency of various stream-flows into the basin. Adding to the injury, oil and gas companies have cut 10,000 miles of channels through the levees over the years.

Louisianans had long pleaded for help without much success from the US government. In the end, however, it took the BP oil disaster to arouse enough public awareness to do something big. Today a massive restoration operation, funded with BP settlement money, is underway that includes numerous state coastal protection and restoration projects, the largest of which is the $2.5 billion Mid-Barataria Sediment Diversion (MBSD) engineering project, conceived by the Louisiana Trustee Implementation Group—a conglomerate of state agencies responsible for restoring Louisiana's natural resources in the wake of the 2010 *Deepwater Horizon* oil spill.

If all goes as planned, the MBSD will punch an artificial 2.5-mile channel through the levees to direct river water into the basin. That should allow massive inputs of nutrient-laden sediment found in the diverted water to boost fertility and even create new land—maybe up to 30,000 acres of wetland over a 50-year period. Another benefit may be that the higher ground created by the silt deposits will also serve as buffers to future hurricanes and storm surges.

Not everyone is hopeful. Critics argue the project is an expensive boondoggle that won't repair the full injury caused by the BP spill. Nor do all scientists like what they see. Some are skeptical that the project will produce as much new land as promoters promise. Others fear adding water to the basin will only cause more loss of land and thus make storm surges worse. A better, and cheaper alternative to channeling the river, argue some critics, is to use dredged-up sentiment from backed-up canals to create any new land.

Meanwhile, many oyster growers, shrimpers, and other commercial fishers have joined the fray, expressing fears that a huge influx of freshwater would alter the salinity in the estuaries and hurt the seafood industry. "About 2,000 bottlenose dolphins, untold numbers of fish, oysters, shrimp and blue crabs, would be destroyed," warns Margie Vicknair-Pray of the Sierra Club Delta Chapter Conservation Program, "not to mention the livelihood of hundreds of families, some who for generations have lived by fishing, crabbing, oystering, hunting, etc., in the waters of Barataria Bay. There are three Indigenous tribal groups that live in the Basin and depend on it for food and income."[15]

Vicknair-Pray and others have also expressed concern that the water diverted into the basin isn't all that fresh and is instead poison. "Because that is what it is," she insists. "There is a 6,300 square mile area known as the Dead Zone off the mouth of the Mississippi River. This is due to the extreme load of toxic chemicals, fertilizers and pesticides from both industry and agriculture along the Mississippi. This project would transfer some of that water into the now productive Barataria Basin, turning this once fertile area into a second Dead Zone."[16] In July 2022 for all these reasons, the Executive Committee of the Delta Chapter of the Sierra Club voted to oppose the MBSD.

Not all environmentalists agree with the club's action. Many, in fact, believe MBSD may succeed in mitigating many of the problems feared by the critics with good designs, and construction. Besides, they add, doing nothing is not smart. "It is our best shot to push back against sea level rise, subsidence and hurricane storm surge, but since our state is already washing away, the sooner we put it back to work, the better," argues James Karst, director of communications and marketing for Restore the Mississippi River Delta, a coalition of environmental organizations.[17]

However, Vicknair-Pray remains unconvinced. "I'd have to say that the future of the Gulf Coast is dicey, especially here in Louisiana where it's been sliced and diced by the oil and gas industry for a century," she says. "On top of the physical destruction, you can add the industry vomit and agricultural wastes and paint a pretty dismal picture for south Louisiana."[18]

Other pretty dismal pictures may soon materialize elsewhere in the South, if climate change remains largely ignored and unchecked. Such negligence is likely to be not only destructive, but expensive, too.

Insurance Blues

Finding insurance for homes and commercial properties, along with flood and wind damage coverage, is already a big headache for many homeowners across the United States; it is becoming tougher to get as major insurers, with their eyes on climate change predictions, deny coverage, raise rates, or even stop offering property coverage in certain states.

The insurance industry, in fact, has long pondered the risks and potential costs from damage from hurricanes, flooding, and other climate-related maladies, and crunched the numbers. Insurers know that properties threatened by flooding in Florida, for instance, are expected to increase 17.5 percent in 30 years. Policymakers also realize 80 percent of the state's population lives near the coast. Small wonder that 15 percent of homeowners in the state in recent years are self-insuring, because they couldn't find affordable coverage for their homes.

In June 2023, Farmers Insurance announced it would no longer write new policies for properties in Florida, citing catastrophic costs resulting from mounting prices in housing, lumber, and labor, along with rising costs of claims. AIG, another insurance juggernaut, has already stopped issuing new policies. Meanwhile scores of similarly bottom-line Southern-minded insurance companies also worried that in addition to other factors such as fraud and litigation costs, the risk of hurricane damage and flooding meant it was time to reduce coverage or pull out of the insurance market in other locations.

Such a trend is extra worrisome for those whose coastal homes lie in areas publicly spotlighted by insurance companies as having the highest flooding risk. According to a May 18, 2023, report from Assurance, an online insurance marketplace (which ceased operations in 2024), southern beach towns make up half of its list of the "20 Riskiest US Towns to Own a Beach House." Florida, Georgia, South and North Carolina made the list. So did Texas, where some Galveston beach homes are now labeled as being in the most jeopardy, with a flood risk that is 58.3 percent higher than the national average. Galveston also has some of the country's most valuable coastal property. The Assurance analysts figure the total value at about $441 million. Such bad publicity makes pulling off a resell a daunting, if not impossible, task.

Insurance woes are hardly confined to the South. Hundreds of cities around the country also have the climate insurance blues. Rising seas are to blame in Atlantic City, New Jersey, but rate hikes for flood insurance have also gone up in Duluth, Cincinnati, Chicago, and New York. Firestorms and earthquakes, along with sea level rise, are to blame in California.

Even if you can get, or afford, insurance to protect these investments, and are able to raise your home on stilts or by some other architectural connivance designed by engineers, would you want to stay put with all that water sloshing underneath the floorboards and inching upward all the time? This is a big worry for homeowners on the barrier islands of North Carolina's Outer Banks, where beaches have been eroding. In May 2020 one house fell into the sea; by October 2025, 20 more had collapsed.

Head Out on the Highway!

Despite these assessments, people continue flocking to the South to live. That may change, predict some experts. "Sea level rise will eventually force people from coastal areas causing a re-ordering of the US population," says Matthew Hauer, an assistant professor of sociology at Florida State University whom many of his peers have deemed a "rising star" in the new field of climate demographics. The coming migration, Hauer predicts, "will be larger than the Great Migration"—the name given to the exodus of millions of African Americans away from the racist and oppressive Jim Crow South. This unfolding event is unique because we know it's going to happen, he says, and "we are equipped to understand how this migration will likely unfold."[19]

Before Hauer began his work, he noticed prevailing climate migration research largely focused on combining coastal inundation mapping with existing population figures to derive an estimated number of people likely to be at risk because of sea level rise. Hauer, however, improved the outlook by considering US Census data, population projections, IRS migration data, and highly detailed sea level rise inundation mapping on a neighborhood level to produce a clearer picture of the risk areas and where those who might be displaced may resettle. His computer models produced estimates of how many people may be forced to retreat, the demographic details of those at risk, and their possible destinations if they can relocate. Using a low-end estimate of a three-foot sea level rise (which is the lowest on a range of possibilities used by scientists), Hauer sees four million Americans at risk; but when he crunches data employing a six-foot sea level rise (which is at the high end of the range) the estimate jumps to 13 million people at risk, mostly in the Southeast, with six million of them in Florida.

Using data from 2020, Hauer estimates over 50 percent of US coastal residents now are white and non-Hispanic; but as the century progresses, he expects the number of residents who are Black, Hispanic, Asian, and biracial are going to grow while the level of the white population will decrease. A dynamic based on the age of residents could emerge, too. For one thing, America is becoming older every day, which means there will be an increase in the number of residents over 55 in many of these coastal communities, and the majority will be women.

Where will climate migrants go? As always, factors such as social networks, economic conditions, and travel expenses all will play a role. Hauer's research provides a tracking of prior migrations that helps in understanding where future destinations might be. At first many climate migrants could be expected to move inland or head to other cities in the South such as Austin, Orlando, Atlanta, and Houston (which Hauer thinks could each see 250,000 new arrivals by 2100). Other major urban areas, meanwhile, will see their populations drop by the tens of thousands. Miami may lose up to 2.5 million people.

But are "destination cities" ready and willing to receive hundreds of thousands of new residents? After all, many municipalities, such as Atlanta, are already facing water shortages, infrastructure problems, and growth management woes. These communities, like others, will also be coping with the vagaries of climate change, whose outcomes nobody can fully foresee.

A sampling of migration yet to come is now available, thanks to the destructive impact of King Tides. These are not rain events, nor storm surges, but rather naturally occurring high tides that typically come ashore twice a year, during spring and fall, when the moon and sun appear together on the same horizon and exert extranormal influence on the tidewaters of Earth. Thus, amplified by the ever-rising seas, King Tides are now so powerful and last so long that they cause significant flooding of urban streets, businesses, and homes along the coast. King Tide flooding, in fact, has become so bad in Miami Beach that city officials have been forced to erect costly, and increasingly ineffective, sea walls and have spent over $400 million for pumps to evacuate the nuisance saltwater from city streets.

Hauer's computer models suggest that about 800,000 Americans are exposed to this kind of annual flooding, and by 2100, there could be "as many as 45 million people living in neighborhoods that flood at least once a year with three feet of sea level rise."[20] His sea level rise migration scenario provides a picture of only part of a broader impact that may yet occur. The increase in the number of days defined by punishing heat and humidity in the American South is also likely to make life miserable for millions who may forsake their home states altogether and head northward or to higher latitudes.

Escaping the Flames

Drought and firestorms, which are likely to affect parts of Texas, Oklahoma, and Arkansas, are likely to become pushers of out-migration. This process has already begun in arid, fire-prone California, where according to the California Department of Forestry and Fire Protection, the state's wildfire season was a record-setting one during 2020 with nearly 10,000 wildfires burned over 4.2 million acres—that's more than 4 percent of the state's 100

An ominous looking dust storm blowing in over a western Oklahoma farm. iStock.com/Susan Vineyard.

million acres of land, making the wildfire season the worst in modern history. The high cost of housing had already been driving many Californians out of the state. More followed in the wake of extreme weather events. A study appearing in the August 24, 2022, edition of the journal *Population and Environment* indicates roughly a third of those Californians sampled said they "intended to move in the next 5 years, nearly a quarter of whom reported that wildfire and smoke impacted their migration decision at least a moderate amount."[21]

Another study, the *2021 Migration Report*, published by North American Moving Services, foresees movement away from the nation's two heavily populated coasts and moving to the Midwest, the Southwest, and South, which includes the top five destinations, the Carolinas, Tennessee, Florida, Arizona, and Texas.[22]

In the future, other forces, such as changes in social and economic conditions and the revamping of zoning and land-use rules, along with the success or failure of adaptation efforts and the unexpected impacts of future climate change management could also affect residents' decisions about relocating. Unchecked sea level rise, for instance, will cover valuable and taxable real estate, depriving residents of their homes and businesses and local governments of revenues. Nobody yet knows to what extent other factors such as diminished freshwater supplies, and the disappearance of jobs as businesses and industries relocate to more favorable climes, may have on migration patterns.

To Leave or Not to Leave

Some people at risk will have the resources to make an early escape; others may follow as conditions deteriorate in their communities. There will most likely be some individuals who for any number of reasons relating to age, wealth, education, mental and physical acuity may elect to remain in place and endure a disproportionate share of climate misery. Demographers already speak about "trapped communities."

Market forces are also at work pushing less fortunate people from their homes, even if they want to stay, as the climate crisis mounts. Developers in many urban areas have already swooped into high-risk zones where they can buy properties cheaply, remodel them to be climate-resilient and turn them into chic-environs for wealthy clients. Some call the process climate gentrification. In Miami, for instance, wealthier residents are abandoning their mansions on the Atlantic coast and relocating several miles inland, gentrifying neighborhoods of the less affluent and pricing those residents out of their homes.

Climate migration, of course, is not unique to the South. It is already underway around the world. Rising global temperatures, the spread of crop diseases, and extreme weather events, such as powerful hurricanes, explain, in part, the flight of thousands of people from Central America to the US border in recent years. In 2018 the World Bank warned that within the next three decades, as many as 143 million people from South Asia, Latin America, and Sub-Saharan Africa could become climate change migrants.

As the seas rise, so do demands for a moral reckoning from a growing number of social justice activists, especially those whose ancestors and current families have suffered a disproportionate share of toxic industrialization, colonialism, war, slavery, and authoritarianism. "As is always the case, the poor and vulnerable are the first to suffer and the worst hit from Climate change," observes UN Secretary-General António Guterres. All too often, this means people of color, Indigenous people, women, children, and the very old.[23]

Nobody can escape climate change, but those who live along seacoasts and in warmer places, whether it be Bangladesh, Mississippi, the Bahamas, or islands in Micronesia (which are now disappearing because of sea level rise), may face the harshest challenges. Extreme weather events, such as ever-more-powerful hurricanes and flooding, are likely to be deadlier and more destructive. Disadvantaged communities along many coastlines must contend with encroaching sea rise and old infrastructure that is subject to power outages and water and sewage issues after storms. Residents here, if they are poor and lack political power, are less likely to receive emergency and disaster assistance as quickly and abundantly as more affluent communities, meaning they suffer disproportionately compared with others. In general, say researchers, this disparity occurs because people of color along with almost anyone who is poor, powerless, or marginalized endure a lopsided share of exposures to toxic waste sites, polluting factories, busy highways, power plants, failing septic tanks, and stormwater retention ponds.

All too often their neighborhoods, whether by design or neglect—were the ones turned into garish industrial landscapes. Or their homes built upon old chemical dumps. Or their wells contaminated with lead and other industrial poisons. Or their sleep disturbed by smelly delivery trucks endlessly passing their homes groaning with loads of chemicals, fabricated products, and waste.

Not only have residents in such places often borne the most severe physical and psychic costs of industrialization, they also historically have received fewer benefits of living in the modern world. Many also lack proper diets and access to healthcare. And they often don't have the resources to fight back in the political and legal arenas to mitigate the harm. The arrival of weather extremes is likely to make living in many of these areas unbearable. "These results of climate change," says FSU's Matt Hauer, "will continue to develop as a source of inequality this century and beyond, transmitting inequality across time and saddling future generations with environmental and social disadvantage."[24]

All of this is simply not fair, argue a growing number of activists around the world. That's why many of them in the South over recent decades have been mobilizing to tackle what they see is a newly understood kind of inequity called environmental injustice.

TWO
FROM THE SOUTH A NEW CIVIL RIGHTS MOVEMENT

The concept of environmental justice appeared in the 1970s and has been gaining ground ever since, especially in the American South. In 2013, according to *Forbes,* the South was home to the nation's top three polluters—American Electric Power, Duke Energy, and Southern Company—which emit a combined 375 million metric tons of CO_2 every year.[1] And by 2022, Louisiana had the dubious honor of being America's worst polluted state, with much of the contamination issuing from more than 150 petrochemical plants and refineries located in an 85-mile stretch of land along the Mississippi River, running from Baton Rouge to New Orleans and slicing through the River Parishes of Louisiana. This ugly, stinking industrial belt, nicknamed "Cancer Alley," also provides employment for the mostly Black residents in the region. By working and living where they do, however, many residents fear they run a higher-than-normal risk of developing cancer and other serious ailments. A 2022 Tulane University study supports those suspicions. Louisiana, concluded the researchers, had the second-highest incidences of cancer in the nation.[2] Life could become unhealthier in Cancer Alley if a Taiwanese-based conglomerate, the Formosa Plastics Group, succeeds in overcoming opposition to its effort to gain approval to build an additional 14 petrochemical plants—a 9.4-billion-dollar project—on the west bank in St. James Parish, Louisiana. At least, that's the fear commonly expressed by local residents.

Similar conditions abound elsewhere in the South. According to Megan Mayhew Bergman, writing in the *Guardian:* "Pahokee, Florida, a town whose population is 56% African American and 29% Hispanic, has had to confront the sugar industry, which polluted nearby Lake Okeechobee, endangering drinking water, fish safety, and property values. Paper mills have polluted Africatown, Alabama. Burlington Industries dumped cancer-causing PCBs in Cheraw, South Carolina. There's toxic coal ash in Uniontown, Alabama."[3]

Moreover, "The south-east, particularly North Carolina and Virginia, is notorious for its coal ash deposits, spills, and anti-regulation mentality," Mayhew Bergman points out. "People of color have outsized exposure to coal ash pollution, which contains carcinogens like mercury, lead, and arsenic. The EPA [Environmental Protection Agency] estimates that 1.5 million people of color live in areas vulnerable to contamination."[4]

Even those who earn their livelihood far from industrial smokestacks may be at extra risk. Consider the Vietnamese immigrants who since the 1980s have been fishing the Gulf coasts of Texas and Louisiana to earn their livelihoods. It hasn't been easy for them or for Cajun, Black, Creole, and Indigenous residents working in waters tainted by industrial pollution from nearby floating oil wells, petrochemical plants, and wetlands swamped by sea rise.

These problems aren't confined to the South. For instance, one study conducted by the EPA-funded Center for Air, Climate, and Energy Solutions, lasting May 1, 2016, through April 30, 2021, found that African Americans, Hispanics, Asians, and other people of color in all income groups across the United States had a 15 percent higher overall exposure to regulated air pollutants called fine particulate matter ($PM_{2.5}$) than white people.[5]

According to the Environmental Defense Fund, 55 percent of Latin Americans live and work in climate-stressed states like California, Texas, and Florida. Since many Latinos work in manufacturing, construction, and agriculture, they encounter intense, prolonged heat and chemical pollution. Along with other residents in low-income communities, they are more likely to be exposed to environmental hazards and experience a gap in the availability of resources, information, and services. At times, not having access to information in languages other than English has made such situations worse during emergencies.[6]

Many marginalized communities overly exposed to the menace of environmental degradation, whether by design or not, are functioning as industrial *sacrifice zones.* Elected officials, often defensively, cite factors such as real estate prices, transportation routes, and land-use designations to explain why a particular factory, landfill, or highway appears in a poor, underprivileged neighborhood instead of a more affluent one where residents have the financial and political resources to challenge, alter, or even halt projects they don't want.

There may be other reasons, too.

Dark and Dangerous Days in Turkey Creek

The residents of Turkey Creek, Mississippi, have solid reasons for thinking they have been living a long time in a sacrifice zone. And it is one they are determined to hold on to.

Today, their historic African American community is a preserved patch of green along a polluted creek, sandwiched between strip malls and the Gulfport International Airport. That it exists at all is a testament to the fighting spirit of Turkey Creek activists who years ago rose up to stop their beloved community from being obliterated for the economic betterment of others.

The community began in the Reconstruction era during the second half of the nineteenth century when formerly enslaved Black Mississippians founded a town along 13-mile-long Turkey Creek, a waterway running through a swampy area north of Gulfport. Over the years, they and their descendants turned buggy, snaky marshes into productive farmlands. Like many other laborers across the South, many of them worked at a factory that produced creosote, an oily, highly flammable substance used as a wood preservative and a probable carcinogen. Despite all hardships, the people of Turkey Creek created a thriving community with its own school, businesses, and church.

Today, those living in the historic community still share a fondness for the creek, no matter how polluted it is. Some even consider the waterway sacred because their pastors baptized the faithful in its muddy, dark waters. Older men and women can also recall when the creek was the only swimming spot available to local Black people who dared not appear on the nearby whites-only beaches on the Gulf coast.

During the 1990s, the nearby city of Gulfport was busy expanding its municipal boundaries. Day by day, the city's infrastructure, roads, and construction projects crept closer to the boundaries of Turkey Creek. At last, the Gulfport City Council, with no input from the residents of Turkey Creek, annexed their low-income town. Next, the council passed new zoning laws paving the way for the construction of outlet stores such as Walmart and Family Dollar, along with a mass of power and telephone lines and treeless asphalt parking lots in the area. By now the residents of Turkey Creek were convinced Gulfport's movers and shakers were banking on everyone living in their community to be tempted by rising land values and sell their land to speculators and move away.

One morning in 2001, without warning, bulldozers appeared not far from the town center and, before anyone could react, demolished an African American graveyard to make way for residential construction. This sent shock waves through Turkey Creek. Fears intensified when residents of the town learned other developers were also on the way, including one with construction plans to fill in wetlands near the creek—a waterway dismissed by one Gulfport city council member as a drainage ditch.

"It's not drainage," insisted Derrick Evans, who gave up a teaching job position at Boston College to return to Turkey Creek, the place of his birth, and lead the grassroots group that arose to defend the creek and the community. "It's about the irreversible consequences of pro-development tunnel vision in a wetland that's historically valuable and endangered."[7]

The developer, Butch Ward, not surprisingly, saw things differently. "I disagree that I'm filling in wetlands because I don't believe there are wetlands there," he insisted. "It's somebody's else's definition of a wetlands."[8]

That somebody else, however, was the EPA. Yes, said the agency, much of the land in question was indeed a wetland—an important piece of real estate on a hurricane riddled coast. That is why Ward needed a permit to pursue his project and why it came to a permanent halt.

Nature Strikes

This wasn't the end of environmental woes for the people of Turkey Creek, however. On the morning of August 29, 2005, Hurricane Katrina made its final landfall on the Mississippi Gulf Coast, slamming the area with 120 mph winds and 28-foot storm surges, leaving some Turkey Creek residents neck-deep in stormwater and much of the community shattered.[9] Making matters worse for storm survivors and other poor people in Mississippi, many still living in Federal Emergency Management Agency (FEMA) trailers, the state's governor, Haley Barbour, in 2008 diverted $570 million of federal recovery dollars intended by Congress to help supply affordable housing as hurricane relief and used the money instead to expand the port of Gulfport. What made this possible was a waiver from the Secretary of Housing and Urban Development. Critics complained that once again "good old boy" politics favored the state's rich and powerful at the expense of the poor.[10] A case study in environmental injustice in Mississippi, they claimed, was in full display.

Still more trouble was on the way. On April 20, 2010, the BP *Deepwater Horizon* oil-drilling rig in the Macondo Prospect in the Gulf of Mexico blew up and sank, killing 11 workers and creating an environmental catastrophe. According to NOAA, the rig spilled 134 million gallons of oil into the Gulf of Mexico, making it the largest offshore oil spill in history and the nation's largest maritime disaster. The catastrophe inflicted lasting physical, economic, and emotional damage to humans, flora and fauna and ecosystems across 1,300 miles of shoreline throughout five states along the Gulf, including the people of Turkey Creek.

Despite all these adversities, Turkey Creek's campaigners, with outside help from the Sierra Club and other organizations, were victorious in getting the town named a historic community district. In 2021, the nonprofit American Rivers designated Turkey Creek among the nation's most endangered rivers.

Today, the Turkey Creek story illustrates how economic development projects can unfold with rapid, machine-like efficiency, damaging not only the environment, but also human well-being. What may be "progress" for some stakeholders is often injury to others, frequently the poor, especially people of color, whose elected officials often fail to consult or care about them because, as the African American novelist, Ralph Ellison wrote, they are "invisible."

That lack of visibility is when injustice begins.

Drying Up in Jackson

Neglect and possible racism can play a role too, as they have allegedly in the city of Jackson, Mississippi—the state capital. Racial tension is frequent in the Magnolia State capital but even so, the old suspicions and tropes were hard to miss in August 2022, when the city's freshwater supply collapsed, leaving 150,000 residents—of whom 80 percent are Black and about one out of four live in poverty—no safe drinking water for weeks. An immediate cause, local city officials said, was that Jackson's main water treatment facility—which was badly in need of maintenance, repair, and improvements—had become overwhelmed by heavier than usual rainfall in August of that year.

Finger-pointing instead of cooperation quickly got underway. The city mayor, Chokwe Antar Lumumba, claimed Jackson lacked the resources, workforce, and expertise to keep the aging and leaking utility running properly and that white flight in recent years was just one reason for the city's shortage of funds. Another, he asserted, was the failure of the state government [mostly white and Republican] to approve bonds the city desperately needed to improve its water infrastructure.

Others saw more sinister motives at work. "We believe we gave compelling evidence that the state of Mississippi intentionally starved the city of Jackson of the resources to maintain its water infrastructure," NAACP President Derrick Johnson, a resident of Jackson, told the Associated Press in October of that year.[11]

Mississippi's Governor Tate Reeves, however, rejected these allegations, and instead publicly blamed Jackson's water woes on Lumumba's incompetence in managing the city's infrastructure operations. This caused a stir. Reeves caused another not quite a month later when he seemingly made fun of the enduring water problems in the state capital. "I've got to tell you it is a great day to be in Hattiesburg," he reportedly told a crowd at an outdoor event in southern Mississippi. "It's also, as always, a great day to *not* be in Jackson. I feel like I should take off my emergency management director hat and leave it in the car and take off my public works director hat and leave it in the car."[12]

Outrage exploded across the media landscape. Not long afterward, the EPA announced it was launching an investigation into whether Mississippi state agencies had racially discriminated against Jackson. By May 2024, the agency found insufficient evidence that this had happened, but the city got some relief when it entered a stipulated order with the US Department of Justice and the EPA which appointed an interim third-party water manager for the water system.

A Community That Had to Sell Its Soul

Race discrimination isn't necessarily the reason environmental injustice arises. Poverty is often at play too. In fact, at times in some places, economic conditions are so dire that desperate people believe they have no choice but to make a devil's bargain. "The South has always been very poor and until recently, almost anything that would get people jobs was looked upon as good and you didn't ask too many questions," argues Michael Thomason, a University of South Alabama history professor. "You'd give them anything you'd have to come here. Give them land. They were desperate."[13]

In 1947, such economic desperation was in full force in Taylor County, a rural area located in Florida's Big Bend. By this time, the local timber industry had devastated local virgin forests which sent the local economy spiraling. Things were so bad that county commissioners went to Tallahassee, the state capital, where they convinced the state legislature to designate the 36-mile-long Fenholloway River that flows through their county and exits into the Gulf of Mexico, as Florida's sole "industrial" river. With strong support from the local community, the commissioners argued a sacrificed river would be a great incentive to attract much-needed industry to their economically distressed county and keep young people from leaving the area.[14] It seemed nearly everyone in Taylor County favored losing a pristine river for the sake of economic security of all. Sadly, they got what they wanted. According to a law approved by state lawmakers on April 23, 1947, crafted exclusively for Taylor County, any municipal corporation, person, or firm operating "any manufacturing or industrial plant in Taylor County, Florida, shall have and is hereby granted the right and is hereby empowered to discharge and deposit sewage, industrial and chemical wastes and effluents, or any of them, into the waters of the Fenholloway River, and the waters of the Gulf of Mexico into which said river flows."[15]

Seven years later the Procter & Gamble company, a major manufacturer of health, personal care, and hygiene products, arrived in Taylor County, built the Foley Cellulose mill, and hired hundreds of people to grind up pine trees into a bleached pulp used for a variety of disposable products, such as toilet paper and tampons. The company, however, also polluted the air in the local community and turned the Fenholloway River into a stinking, crow-black sewer drenched in dioxin contamination which according to a 1990 EPA study was 1,900 times acceptable levels.[16] In the years to come, cases of cancer and leukemia and other disorders in Taylor County soared, as residents feuded with one another repeatedly over the deal their community had made.

Over the years, grave concerns about quality of life and job security loomed like a huge shadow over Taylor County. The most recent mill owner, Georgia Pacific, reportedly spent $300 million on cleanup efforts as the community struggled to seek a balance between health and employment.

That search ended when Taylor County was hit by Hurricane Idalia in the summer of 2023, and Georgia Pacific shuttered the mill forever, ending employment for 532 employees and hollowing out the local economy. Adding to everyone's distress, two more hurricanes came ashore in 2024. But it probably brings little solace to those living in one of Florida's poorest counties to know they aren't the only ones who live in close proximity to hazardous waste.

Living and Dying at the Neighborhood Landfill

Many of these other unfortunates live near sacrifice zones known as landfills. They also tend to be poor. In fact, as many as 70 percent of the contaminated waste sites in the United States exist near low-income housing, according to a December 22, 2017, report from the Associated Press (AP). Moreover, an estimated two million people live within a mile of one of the nation's 327 superfund sites, making them more vulnerable to climate change–related flooding, and to rising groundwater. In most of the areas, says the AP, are low-income communities and communities of color.[17] As of 2024, 1,340 superfund sites are spread across the United States; roughly one-fourth of these are in the South.

Nobody wants to live near landfills, which are stinking places, teeming with garbage, toxins, and rats. What came before them, however, was even worse. Garbage dumps, as they were once called, were typically open pits filled with waste and trash, often gobbled by vultures or free-roaming pigs, or in some cases burned.

Around the mid-twentieth century, growing concerns over health, sanitation, and the environment led to replacing a dump with a modern landfill—an airtight tomb designed by engineers and scientists and lined with clay and other nonporous barriers to keep the sealed-up contents from leaking toxic substance and gases such as methane, a greenhouse gas. No matter how tough the EPA requirements for landfills, most people still want to avoid them. It's no secret that container liners used to contain the messes do not last forever, and that cracks can occur and allow the release of dangerous chemicals and contaminants. The surface of the earth and the water below and above are always moving, too, altering landscapes and what lies beneath.

Landfills, unfortunately, are indispensable necessities. But where do you put them? Deciding who gets stuck living near one has long been contentious, if not explosive, in the South, where cheap land, low taxes, sparsely populated rural towns, and unincorporated areas with lax environmental regulations, zoning laws, and ordinances have made the region attractive to faraway people in search of a dumping ground.

Tragedy Launches a New Civil Rights Campaign

It was at one of those southern landfills near Houston, Texas, more than 60 years ago that a tragedy involving a young boy seized the nation's attention and sparked an early attempt at achieving environmental justice.

Houston at the time had no city-wide zoning laws. (Today it still doesn't.) For several years, the city had struggled to keep up with the growing amount of garbage being generated by a rising influx of new residents pouring into the city, one that was fast becoming what local boosters bragged was "The Oil and Gas Capital of the World." In fact, Houston's population grew from about 140,000 residents in 1920 to more than one million 50 years later.

To manage the growing tons of garbage, city officials created another landfill on the urban outskirts in a historically Black neighborhood called Sunnyside. Why there? The residents of Sunnyside were used to having an existing nearby landfill, so how could one more make any difference? (At least, critics of the idea surmised this was the logic.) Thus, by the early 1960s Sunnyside had become home to two landfills and an incinerator.

The residents at Sunnyside, however, already weren't happy. Now they grew angrier and more resentful because it looked like their community was doomed to receive even more garbage and waste, much of it from people who lived far away. No matter how much they complained to Houston authorities about how unfair this all was, they received little attention. Instead, trucks kept arriving to unload stinking debris in their neighborhood, day after day.

On the afternoon of May 8, 1967, Victor Roy George, an African American 11-year-old boy, along with two of his fifth-grade playmates, rode home at the end of the school day. On the way, they parked their bicycles at the Home Road Landfill in Sunnyside—or rather a "garbage dump" as it was described later in court—as they had often done before. This visit to a nasty, dangerous playground for poor kids, however, proved fatal when Victor, trying to fetch a piece of Styrofoam floating with a stick, accidently fell into a hole of polluted muck, 12 feet deep and 30 feet wide, from which he couldn't escape and died.[18]

News of the boy's death sped through "Black Houston," as it was then called. As so many other places across the South and the nation, the community was already fraught with racial tension over ongoing civil rights struggles. So, it didn't take long for angry protesters to rapidly mobilize to express furious indignation, grief, and sympathy for Victor George. Their most dramatic act was to stage a sit-in at the entrance of the long-hated dump to prevent more trucks from making deliveries.

Others soon rallied in support at hastily arranged rallies at local churches and other public places, including a nearby university where during the night discontent escalated into a violent clash between students and local police—a common turn of events at the time. What made this protest unique, however, was that perhaps for the first time demonstrators were demanding something that later received a name—environmental justice.

Other than making history, however, the Sunnyside protests accomplished little. The landfills remained open. Victor's parents later sued the city government but lost.

However, the battle cry of environmental injustice burst into public discourse and would soon show it had staying power. Moreover, Houston's landfill selection policies were still unfinished business. They came under intense scrutiny again in the late 1970s, when a local environmental movement morphed into something bigger and more potent.

Taking the Fight to Another Level

As Robert Bullard, a distinguished professor at Texas Southern University, remembers it, his wife, Linda McKeever, an attorney, one day returned to their home in Houston after work and said, "Bob, I've just sued the state of Texas, Houston, and Harris County, because this company is trying to locate a landfill in the middle of this predominantly Black middle class suburban neighborhood in northeast Houston."[19]

She also informed him she urgently needed somebody to do a comprehensive study including a map showing "where all the landfills, incinerators and garbage dumps are." That somebody she had in mind was her husband. He was a smart choice. As a trained sociologist, Bullard knew how to do a proper study that could command attention, especially in court.[20]

He agreed and got to work. What Bullard soon learned was that Sunnyside was no exception when it came to being selected by a local government in the South as a garbage disposal site. In fact, his research revealed that five of Houston's existing landfills, six of its public incinerators, and three of four privately owned landfills in the city had been set up in predominantly Black neighborhoods in recent decades. That meant Houston's Black population, which made up 25 percent of the city's total, was disproportionately receiving 82 percent of the urban garbage and waste.

This information supplied evidence for the legal work Linda McKeever was doing. In 1979, she filed a lawsuit, *Bean v. Southwestern Waste Management,* seeking a court injunction to halt construction of the new landfill. The suit also alleged racial prejudice was behind the decision to construct a landfill in the predominantly Black community, which was a violation of new federal civil rights laws. Never until now had anyone made a claim like this one in an American courtroom.[21]

The dispute went to trial at the US District Court for the Southern District of Texas in Houston. On the opening day, it was clear that a racial chasm existed in the makeup of the key players in the trial. The plaintiffs, their attorneys, and expert witnesses were Black. But the defense lawyers and the judge were white. Setting many teeth on edge, as Bullard remembers it, was the trial judge who kept referring to the Black plaintiffs and the residents of Houston's African American communities as *Nigras.* "Anyone growing up in the South knows that 'Niggra' is another way some southerners call black people the other 'N' word," Bullard recalled later.[22]

The insensitive judge, however, did agree the plaintiffs had proven something substantial; this he made clear in his written decision describing why he was ruling *against* them: "The plaintiffs have adequately established that there is a substantial threat of irreparable injury."[23]

The problem, he added, was that "The burden on them is to prove discriminatory purpose." And this, said the judge, they had not done.[24]

Thus, it was not enough to prove local governments had dumped on Black communities disproportionately. You also had to convincingly prove with a preponderance of evidence—that is, a legal demonstration that something is more likely to be true than not—that the defendants had used *race* as a basis for their acts of discrimination, something hard to do. Intent, not the outcome, was what mattered to the judge.

Midnight Dumpers

What might be hard to prove in court, however, did not seem hard to grasp for thousands of residents of Warren County, North Carolina, more than 1,200 miles away, whose population was 65 percent Black, a fact which many of them believed was the reason state authorities had decided to dump 60,000 tons of PCB-contaminated soil in their community.

PCBS, or polychlorinated biphenyls, are industrial products or chemicals used in fabricating insulation for electric products and thought to be linked to cancer and other human health problems. The PCBS at the heart of the Warren County dispute came from the Raleigh-based Ward Transformer Company, a firm specializing in rebuilding used voltage transformers, which at the time contained oil-laced PCBS. By 1978, the federal government had tightened regulations on the handling and disposal of these carcinogenic fluids, which made them undesirable for manufacturing purposes, as far as Robert Earl Ward, chairman of the board for the company, was concerned. That summer he contracted with his friend Robert J. Burns, owner of a trucking company, to haul the cancer-causing fluids away from his manufacturing site to a special recycling facility in Philadelphia, Pennsylvania.[25]

Within a few months, however, Burns realized his venture was a moneyloser. So, he decided to take a shortcut, an illegal one. Under the cover of darkness, he drove a truck equipped with a 750-gallon tank and dumped contaminated oil from Ward's factory onto federal property at Fort Bragg, a military reservation in North Carolina.

This was obviously a cheaper way to conduct his mission, but it was illegal, cumbersome, and left a big telltale toxic mess. So, Burns and his two sons tried yet another disposal method. This time they spent several nights, using a special hose connected to the tank in their truck, to spray an added 30,000 gallons of toxic oil onto the ground along 240 miles of rural roadsides in 14 North Carolina counties. These "Midnight Dumpers" thought disposing smaller amounts of oil over a longer stretch of land would be easier to get away with.

They were wrong. Authorities had little trouble catching up with them thanks to 240 miles of evidence. A trial took place, a jury convicted Burns and Ward, and a federal judge sentenced the pair to prison for illegal dumping. Burns's sons received suspended sentences after agreeing to testify against the Ward Transformer Company.[26]

Perhaps the criminal justice system was satisfied, but there were still tons of contaminated North Carolina roadside dirt to contend with. For the next four years, Governor James B. Hall and the state government saw to it that workers posted yellow warning signs on the contaminated areas and spread sand atop the tainted roadsides. None of this, however, controlled the spread of the contamination, as it seeped through the ground. This worried many North Carolinians in the polluted areas, alarmed by the presence of a vast three-foot swath of stinky, greasy, brown goop along the roadsides. Was there a danger of toxins in the air? What if the dreaded PCBS spread into crops, waterways, and animals? And wasn't there an immediate danger to school kids who might ignore warning signs and traipse through PCB-drenched turf?

After a long period of dickering with the EPA about disposal methods and considering various soil and water samples taken from at least 90 potential sites in 14 North Carolina counties, the governor decided the best thing to do was to scrape up all the contaminated roadside soil—all 40,000 cubic feet of it from the 14 counties—and bury the whole mess in just one county—poor, sparsely populated and mostly Black Warren County, located on the Virginia border and north of Raleigh. The doomed tract of earth was an existing 142-acre landfill found near the rural, unincorporated community of Ashton, North Carolina. It was the safest place in the state for such an undertaking as this, said the governor, and served the public's interest, too.

It was also far cheaper to dump it all in Ashton than to transfer the nasty stuff to incinerators or send it out of state. EPA and state officials claimed the designated 22-acre hazardous materials pit would be built correctly with modern designs. No PCB-tainted water would ever escape the "dry tomb" they were planning for encasing this deadly stuff. Why, Ashton would be receiving the "Cadillac of landfills."[27]

This assurance did not placate the alarmed citizens of Warren County. They knew you didn't need to be a scientist to sense something wasn't right when North Carolina's "safest site" did not receive rigorous soil and analysis until months *after* Hunt's decision. That is when state officials and the EPA concluded the site was not proper for toxic waste disposal. The soil was not compact enough, they said. The water table was too high; in fact, all you had to do was dig down five feet and you would strike water.

But these findings created no lasting obstacles for the state. The EPA simply granted waivers to get the job done and suggested ways to avoid environmental damage.

An Aroused Citizenry Reacts

Meanwhile, 30-year-old Dollie Burwell, the child of sharecroppers and a veteran civil rights activist, along with many other Warren County residents, including a white couple—Debra and Ken Furrucio, who had connections to environmental groups—organized local resistance to the toxic apocalypse they believed was coming their way.[28] That they had done a respectable job of stirring things up became clear when hundreds of people showed up at several public meetings in Warren County during the next three years. These people wanted answers. Why weren't they consulted beforehand that the state of North Carolina had targeted their county to receive the contaminated soil? Why were those waivers granted? And was anyone listening to the expert hired by the protesters?

The local activists also had support from outsiders, including William Sanjour, the chief of EPA's Hazardous Waste Implementation branch. "Landfilling is cheap," he said at one rally in the town of Warrenton. "It is cheaper than the alternative. The people who like to use landfills such as chemical industries are very powerful. No amount of science, truth, knowledge or facts goes into making this decision. It is purely a political decision. What they listen to is pressure."[29]

Stopping a state project blessed by the governor seemed daunting to the mostly rural residents of Warren County. But to their surprise and delight, their resistance efforts were making headlines. They were also forging something rare in the South in those days: a biracial alliance. African Americans and whites had showed up together to speak out against what they saw as a common threat. Deborah Ferrucio later told *Climate News:* "From the beginning there was no racial conflict. We all knew that PCBS are color-blind and they were going to affect us all." Previously, she said, the people of Warren County had never faced an issue that united them. But opposition to the landfill "brought them together in a way where black and white people were in the churches singing and holding hands, they were in front of the capitol singing and holding hands." There were even groups, she recalled, as diverse as the United Daughters of the Confederacy and the NAACP that took part in a "new melding of people."[30]

The NAACP's help took the form of a lawsuit aimed at stopping the landfill project. Warren County, the civil rights organization argued, was targeted because its residents were mostly Black and impoverished. Agreement with this notion was widespread in the local community. "I learned why we were targeted," resident Almena Myles told the *Guardian*. "They chose us because we were rural and poor and they thought we couldn't fight it. They thought we wouldn't understand. It was a crash course in advocacy. We felt we had stepped back in time, like it was the 1960s all over again and we had to fight for our rights as if it was the civil rights movement."[31]

State officials, however, denied that the racial makeup of Warren County influenced their decision in any way. District Court Judge W. Earl Britt, who ultimately decided the civil case, declared, "there is not one shred of evidence that race has at any time been a motivating factor for any decision taken by any official—state, federal, or local—in this long saga." Instead, he said, the state had only considered factors such as soil type, low population density, and available space.[32]

Through it all, Governor Jim Hunt was steadfast in keeping with his decision. The state's disposal plans would continue, he said, "regardless of public sentiment."[33]

In the early months of 1982, the fears of many Warren County residents intensified when the director of the EPA's superfund—a federally funded

The 1982 PCB landfill protest in Warren County, North Carolina, that launched the environmental justice movement in the United States. Photo courtesy of *Raleigh News & Observer.*

deep-pocket used to pay for cleaning up the nation's worst toxic zones—distributed $2.5 million to create a superfund depository in Warren County. This meant that outside experts had ranked the mess headed toward Ashton among the worst hazards in America.

The Trucks Arrive

At last, on September 16, 1982, dump trucks carrying six tons of contaminated soil arrived at the site, making the first of an estimated 7,000 future visits. Waiting for them were hundreds of demonstrators. Most were African Americans, but whites and Native Americans showed up, too. They marched en masse to the site under the intense gaze of 60 North Carolina state police decked out in riot gear. A National Guard helicopter hovered above them.

Hour after hour, they chanted and waved homemade signs with messages such as, "Dump Hunt in the Dump!" Some defiantly stood in front of the arriving dump trucks or lay on the ground to block their entry to the site. A few protesters had come from far away, including Walter Fauntroy, a delegate to the US House of Representatives, whom police arrested after he refused to obey their orders to allow the trucks entry. He was hardly the only one to show defiance. "If you do not desist this unlawful act," a white patrol officer told the crowd, "you will be arrested."[34]

After many demonstrators ignored the warnings, patrol officers escorted them to a jail-bus parked not far away. Law enforcement officers also forcibly relocated others away from the road to clear the way for trucks.

History was made that day. Until now police had never detained civilly disobedient Americans in a landfill. "[People] were up in arms and protesting and blocking the roads and lying down in the highway," recalls one community member. "My wife was in it. But I couldn't join her. We were on opposite sides. She was protesting. I was with the Department of Corrections. I had to come and pick her up and take her to jail."[35]

By sunset, all the trucks scheduled for deliveries that day had successfully made their dumps. The protesters, however, did not lose heart. In fact, in the coming weeks, many returned to the site day after day to continue the protest, causing even more run-ins with law enforcement. During just the first two weeks, police made 414 arrests.

At last, a month after the protest had begun, haulers dumped the last of their toxic cargoes. The fight appeared to be over. The state government had triumphed.

Birth of a Movement

However, the protest had not been in vain. The citizens in Warren County achieved something significant: that is, they had successfully created a protracted, organized mobilization against environmental injustice in the South. Following the path cleared by fellow activists in Houston, they had invigorated and helped to define a new and burgeoning civil rights issue, one long unrecognized and long ignored. No longer could the nation avoid peering at environmental issues through the lens of race, class, caste, and power. Activists were now including words such as "environmental racism" and "environmental justice" in the ever-growing discussion about protecting the natural world. Many were also urging the creation of a new union between two of the nation's most important ongoing crusades, the environmental movement and the civil rights movement.

They were not alone. Social justice activists and environmentalists around the world had been watching what had happened in Warren County and Houston. Many had been fighting environmental justice battles in their own backyards and now felt solidarity and common purpose with the Southerners in Warren County.

Any optimism, however, was tempered by a shared sense of urgency as it became clearer that climate change was worsening existing environmental problems and making it harder to get justice. That's why many activists began agreeing it was high time to join forces and to do something. A place in the South might be ideal for everyone to meet, rally, and figure it out.

THREE
THE MOVEMENT GROWS

Just one year after the last dump truck taking part in the PCB mission rumbled away from Warren County, the US General Accounting Office (GAO) released a survey suggesting that what was happening to the residents living there was hardly unique. According to its findings of four hazardous waste landfills in the EPA's southeast Region IV, "Blacks make up the majority of the population in three of the four communities where the landfills are located. At least 26 percent of the population in all four communities have income below the poverty level and most of this population is Black."[1] News like this electrified activists in the environmental justice movement, for now they had empirical evidence to bolster their claims.

In 1987, the United Church of Christ (UCC) released its own study, commissioned by one of its ministers, Reverend Benjamin Chavis, a North Carolina civil rights activist who, along with many of his congregation, took part in the Warren County protests. Unlike the EPA study, the UCC's provided a national perspective. Among its conclusions was a finding that the most powerful determining variable guiding human decision-making in selecting a site for unwanted landfills wasn't geography, price, or socioeconomic status. Instead, it was race.

"Three out of the five largest commercial hazardous waste landfills in the United States are located in predominantly Black or Hispanic communities," according to the study. "The three landfills accounted for 40 percent of the total estimated commercial landfill capacity in the nation."[2]

One of most infamous was on a 300-acre tract near the town of Emelle in the center of Sumter County, Alabama, one of the poorest counties in the

United States, whose population was about 70 percent Black. The landfill was also America's largest toxic disposal area, one that received—and still does receive—delivery of hazardous waste from 48 states and three foreign countries. In fact, North Carolina state officials had once considered Emelle as a possible site for the contaminated mess Robert J. Burns had strewn upon North Carolina roads.

Alabama is hardly the only state with lots of landfills. In 2018, there were an estimated 1,250 of them across the country, and the region with the most was the South, with 491. Next in line was the Midwest, with 345.

Not long after the UCC report made headlines, landfills played a big role in other environmental justice studies popping up across the country, thanks to a small but growing number of scholars exploring the link between race and environmental danger. One of the most influential was sociologist Robert Bullard, who, following his work on Houston landfills, continued his environmental justice work, eventually establishing a reputation as a pioneer in the field. In 1990, his standing grew even larger when he published *Dumping in Dixie: Race, Class, and Environmental Quality.* Today considered a classic in environmental justice studies, the book chronicles five African American communities in the South, including Emelle-Sumter County, whose residents had waged their own justice campaigns for cleaner and safer environments. Widely considered the first major book of its kind, *Dumping in Dixie* helped energize a burgeoning mass movement.

In January of 1990, researchers at the University of Michigan (UM) sponsored a first-ever conference where fellow scholars gathered, compared notes, and further legitimized the new and growing field. Not long afterward, two of the conferees, Bunyan Bryant, the first African American on the UM faculty, his colleague Paul Mohai, along with academicians and activists, began meeting periodically with EPA officials to urge them to expand the agency's work to include the goals of environmental justice.

The Federal Government Joins

EPA officials listened and then acted. This became clear when the agency for the first time featured environmental justice in the March/April 1992 issue of the agency's periodical, *The Journal.* "Is it time we broadened the definition of environmental protection?" wondered John Heritage on the editor's page. "The physical environment of America's minorities—Hispanics, Native Americans, Asians, African Americans, the poor of any color—has in one way or another been left out of the environmental cleanup of the past two decades."[3]

This same issue also ran an article by scholars Bryant and Mohai which summed up the big idea they'd taken away from the Michigan conference: "For the past 15 years or so, the civil rights movement has faltered . . . but a resurgence of that movement may be taking place as people of color and oppressed communities across the nation begin to redefine their struggle in terms of a safe and clean environment for all."[4]

William K. Reilly, the EPA administrator at the time, stated in that seminal issue that his agency had already entered the environmental justice movement. Readers should know, he wrote, that the collection of data on environmental equity issues was underway. The agency was also hosting workshops on a wide range of equity policies that could be implemented in the government to address the unjust impacts of pollution. The EPA, Reilly promised, would hire more minorities and consult a greater number of minority academic institutions doing environmental justice research. He also gave assurance that remedial action was underway with more to come. New standards would be set for landfills, for example. And yes, the agency had its eye on the impact of lead on poor communities. Californian migrant farm workers could expect the EPA to make sure they had clean water to drink.

By September 1993, the agency had a new advisory committee, the National Environmental Justice Advisory Council, tasked with providing advice and recommendations on a wide range of environmental justice issues. Arguably, these were modest first steps, but they signaled that the EPA was now including the notion of justice in its policies. "The early 1990s was when environmental justice really hit the national radar screen for the EPA and the federal government," recalls Bunyan Bryant. "I believe that the 1990 conference was the main catalyst."[5]

An International Gathering

In October 1991, another history-making environmental justice conference took place in Washington, DC—the first National People of Color Environmental Leadership Summit. Convened by the United Church of Christ Commission for Racial Justice, the gathering was, recalls Robert Bullard, "probably the most important single event in the movement's history."[6]

The four-day event drew activists, researchers, and scholars from all 50 states, Puerto Rico, Canada, South and Central Americas, and the Marshall Islands, most of them people of color. The press was there too, although there is evidence some delegates were wary of engaging with white-led media outlets.

A rapt audience heard one guest speaker after another report the grim environmental conditions related to air pollution, hazardous waste sites, and uranium exposures their respective local low-income and often marginalized communities were confronting back home. Industrial pollution was so bad in the century-old Black community of Reveilletown, explained a Louisiana delegate, that every resident had to be relocated permanently.

There was also a disturbing report from Texas at the Carver Terrace subdivision in Texarkana where homeowners had learned they were living atop a buried, toxic, creosote production plant. For some residents, this news confirmed their growing suspicions that something wasn't right in their neighborhoods. Why, for instance, were there so many unexplained rashes, rare cancers, and multiple miscarriages in Carver Terrace? Why did pets keep dying?[7]

Racism was suspected, too. After all, 20 years earlier, during the Jim Crow era, Carver Terrace had been marketed and touted as affordable housing for Black professionals. What lay underneath the grassy lawns was not revealed to potential buyers at the time, however.

A speaker representing the Havasupai Nation of Arizona took the podium to explain why her people worried about health risks associated with ongoing uranium mining in the Grand Canyon. And Mililani Trask, a native Hawaiian attorney, told of a widespread and persistent fear that had seized her community that stemmed from a reportedly high number of leukemia and other cancer cases among the children on the island of Oahu. "We believe," she said, "this is due to electro-magnetic contamination."[8]

Senator Jeton Anjain, a senator in the Marshall Islands legislature, related how he had long struggled on behalf of his people "to find the truth of the question we have been asking: 'Is our homeland safe?'" Anjain was referring to the contaminated, radioactive aftermath of the hydrogen bomb the US government exploded on the Bikini Atoll in the Marshall Islands in 1954. Was there a link, the senator asked, between the radiation fallout caused by the bomb and high rates of cancer, thyroid disease, and birth defects among the inhabitants living on islands near the bombsite?[9]

A recurring complaint during the conference was that communities of color, no matter where they lived, felt excluded from the decision-making process concerning local environmental issues. Instead, the dominant voices seemed always to be those of commerce, governments, and mainstream environmental organizations.

It was understandable that such perceptions could lead to frustration and anger. But some delegates fretted that discontent, if not addressed diplomatically, could lead to division rather than bridge-building with authorities and mainstream and mostly white environmentalists. Dana Alston, a scholar and one of the founders of the conference, however, tried to discourage finger-pointing and instead stressed the importance of finding common cause. "People of color gathered not in reaction to the environmental movement," she recalled in her written account of the conference, "but rather to reaffirm their traditional connection to and respect for the natural world, and to speak for themselves on some of the most critical issues of our times."[10]

Alston also offered advice on how to build a mass movement. For starters, they should remember the environment was not a stand-alone issue. Rather, it "is woven into an overall framework and understanding of social, racial, and economic justice."[11]

The summit ended with a shared feeling of success. For one thing, it generated widespread publicity and news coverage. Delegates also forged new alliances. They called for more scholarly papers, and pledged to adhere to setting up a list of 17 principles to guide their movement into the future to "ensure environmental justice; to promote economic alternatives, which would contribute to the development of environmentally safe livelihoods; and to secure our political, economic and cultural liberation that has been denied for over 500 years of colonization and oppression, resulting in the poisoning of our communities and land and the genocide of our peoples."[12]

Many of the attendees also found themselves buoyed by a refreshing sense of empowerment emanating from the conference. Some picked up new ideas they could use back home. Activist Tom Goldtooth, who had long tussled with the EPA over environmental problems affecting Ojibway people in Minnesota, experienced a kind of enlightenment when he heard others use terms such as environmental racism and environmental injustice. "This is what I'm experiencing here with the work I'm doing," he recalled later. "Racism rears its ugly head when it comes to protecting Native people."[13]

Perhaps the biggest takeaway from the summit, suggests Robert Bullard, was that it "broadened the environmental justice movement beyond its early anti-toxics focus to include issues of public health, worker safety, land use, transportation, housing, resource allocation, and community empowerment."[14] And as is the case with all social movements, diverse opinions emerged over the movement's purpose and destiny. Analyses and clarifications of injustice issues grew more sophisticated, as scholars, activists, and lawmakers sought ways to better understand and agree on what the core principles of environmental justice really were and what the proper and achievable goals of such a movement should be.

New Voices

Even before the big summit, groups such as the Indigenous Environmental Network, the Southwest Network for Environmental and Economic Justice, and the Southern Organizing Committee were doing environmental justice work. Now they were joined by new organizations, such as the Bullock Center for Environmental and Climate Justice at Texas Southern University in Houston—founded by Robert Bullock—where scholars concentrated on education, public policy analysis, and research-based advocacy for environmental justice projects. In 1992, sociologist Dr. Beverly Wright, who grew up in Louisiana's Cancer Alley, started the Deep South Center for Environmental Justice at Dillard University in New Orleans to provide a resource center that today still offers expertise and resources for environmental justice research projects, especially in vulnerable communities along the Gulf Coast and the Mississippi River's "Chemical Corridor," as well as training for environmental workers in fields such as health and safety.

In 1988, local activists in Harlem, New York, started West Harlem Environmental Action to tackle local sewage, traffic, and pollution problems. Two years later Bemidji, Minnesota, activists representing various Indigenous communities and tribal governments were building the Indigenous Environmental Network to address both environmental and economic justice issues that plagued their communities.

Asian Americans and Pacific Islanders joined the movement too. In 1993, for example, the Oakland, California–based Asian Pacific Environmental Network (APEN), whose membership relies heavily on Laotian refugees and Chinese immigrants, set up operations to help local communities transition away from "an extractive economy based on profit and pollution and toward local, healthy, and life-sustaining economies that benefit everyone."[15] Among other things, the organization works with its members, often in their native languages and dialects, to help them navigate ways to address their local environmental problems. In the Midwest, Chicago Asian Americans for Environmental Justice formed to make sure Asian American groups in America's third-largest city had a greater say in the mainstream environmental movement. Such help is often a critical first step in environmental justice matters, especially for groups that otherwise feel marginalized. In recent years, for example, a growing number of Asian American leaders, activists, journalists, and academics have complained that they are all too often overlooked by major environmental and environmental justice organizations and much of the mainstream media, academia, and government. An oft-cited and unappreciated excuse, they say they hear much too often, is that Asian Americans, Native Hawaiians, and Pacific Islanders (ANHPI) as a group represent only 6.2 percent of the US population, a figure far less than that of other racial groups, which means they do get lost in the shuffle.

Some researchers also fault Model Minority stereotyping, a bias that leads to incorrect conclusions. One misconception, for instance, is that a great many members of ANHPI communities are high achievers who live prosperous, professional lives and generally don't face the environmental injustice realities other people of color do. Such thinking may even convince some environmental justice researchers they don't need to collect data on Asian Americans. It may also explain, complain some Asian American academics, why their own grant proposals to include Asian American in climate justice studies are often rejected.

"The assumption is that all 20 million of us (and growing) are successful doctors, lawyers, or engineers living the 'American dream'—surely no Asian American lives on the frontlines," writes Korean American Christina Choi, the content management and digital production director of the National Resources Defense Council.[16]

Deciding who is eligible for consideration and assistance is just one of the perplexing questions that emerged as the concept of environmental justice morphed into a new interdisciplinary social science that incorporates a wide and complex variety of scholarly theories about the relationships of people with nature in terms of justice, law, ethics, economics, and health. There were many more. How, for instance, do race, gender, and class issues factor into the *worldwide* human struggles over the harmful consequences of hazardous waste, resource extraction, and land appropriations ranging from agriculture, land grabs, and set-asides for nature conservation? Can government and corporate policies lead to unjust environmental outcomes even if they are unintentional? Are efforts purportedly designed to balance economic opportunity and environmental degradation being decided by a wide spectrum of citizens or exclusively by those at the pinnacle of hierarchies with personal vested interests? Does capitalism inevitably lead to environmental injustice? Is collectivism even worse? Does *reproductive injustice* emerge in sacrifice zones where harmful elevated levels of toxic waste and pollution endanger the formation of healthy offspring? Who is responsible morally and financially for making things better or preventing further harm?

Such questions were vexing enough for anyone struggling with environmental justice. Then came another concept that caused new headaches, the notion of *climate justice.*

Climate Justice Arrives

The term climate justice may be a relative newcomer, but as a notion it has been around in climate change discussion for decades. As far back as December 1990, for instance, the United Nations Intergovernmental Negotiating Committee was tasked with drafting international principles to help nations sync climate actions with sustainable development. During their discussions, questions over the meaning of words such as *responsibility, fairness,* and *equity* kept popping up. You can see this in Article 3.1 of the 1992 United Nations Framework Convention on Climate Change: "The Parties should protect the climate system *for* the benefit of present and future generations of humankind, on the basis of equity and in accordance with their common but differentiated responsibilities and respective capabilities. Accordingly, the developed country Parties should take the lead in combating climate change and the adverse effects thereof."[17]

Louisiana native and attorney Colette Pichon Battle has spent time on semantics, too. Since 2005, she has collaborated with elected officials and national funders to find ways to supply equitable reparations for communities hit by natural and human-caused disasters. Along the way, she realized *environmental justice* and *climate justice* weren't the same. The former, she suggests, is mostly about the pollution of water, soil, and air and its harmful impact on human beings and the natural world. *Climate justice,* however, in her view, demands deeper thinking. "Climate change is the most horrible symptom of a system that has been built for a few to extract every precious value out of the planet and its people from our natural resources to the fruits of our human labor," she argues. "This system has created the [climate] crisis."[18]

The two problems are linked. "To achieve one is to work toward achieving the other," Pichon Battle says. "They work together and move together."[19]

It's easy to forget, of course, that everybody, not just fossil fuel companies, is responsible for the climate mess. Anyone who rides in a car or a city bus, wears polyester clothes, or drinks from a plastic water bottle inescapably supports the burning of fossil fuels. Even the processes used to make batteries for electric vehicles also emit lots of GHGS.

Moreover, industrial output of the modern age also provides employment, healthcare, food, shelter, and military defense for billions of people. Those industrial nations mostly in the West—often maligned by social justice activists for the sins of imperialism, colonialism, and slavery—also produced vaccines and medicines, an agricultural revolution, life-saving medical, communication, and transportation technologies, and new green energy systems—that have saved and improved human lives around the world.

Such grand human achievements, however, as Pichon Battle points out, came with a price. The modern, consumer-based economy is tied inexorably to an insatiable consumption of natural resources and expectations of never-ending economic growth. It also gave rise to a society marred by extreme inequality, especially for the nation's minorities.

Such issues were being discussed with increasing vigor and ever-growing lists of data as the climate justice movement matured in the United States and abroad. By 2000, there was finally enough common ground and interest worldwide to justify convening the first-ever international Climate Justice Summit that same year, which took place in The Hague, a city on the western coast of the Netherlands. Organizers purposely timed their event to occur during the same week representatives of the developed world were also in The Hague to attend another big conference, the United Nation's Sixth Session of Conference of the Parties (COP 6), where topics such as setting up compliance details for previous GHG emissions reductions discussion, technology transfer systems, and market solutions for climate change topped the agenda. What those attending the Climate Justice Summit in The Hague's Concordia Theater wanted to discuss, however, was different in tone, purpose, and content. For the most part, they had come to complain, blame, and argue about topics such as climate equity, workers' rights, and justice.

Thus, this alternative conference in The Hague served as a splendid free-wheeling platform for grassroots activists, especially those from developing nations. In a small theater sat some 500 people from around the planet—environmentalists, justice activists, and those who had long been tracking and opposing the polluting practices of corporates, especially those connected to Big Oil. In a manner reminiscent of the earlier gathering of justice activists in Washington, DC, speakers shared troubling stories about environmental inequities in their own homelands and communities.[20]

One of them told of the resistance of the Indigenous Ogoni people in Nigeria to Shell oil-drilling operations in the Niger Delta. This story resonated with a veteran campaigner from Louisiana's Cancer Alley—who then shared her own experience and may have stolen the spotlight. In fact, according to the San Francisco Bay Area–based corporate watchdog organization, CorpWatch: "Perhaps the most dramatic moment in the Climate Justice Summit came when local Louisiana activist Margie Richard confronted Shell executives. Her group, Concerned Citizens of Norco, has been fighting a Shell refinery for years. Richard entered a meeting cohosted by Shell on the Clean Development Mechanism inside the official conference center. As a Shell representative expounded on how the mechanism 'must reflect commercial reality,' Richard gave him a dose of on the ground reality. She presented the surprised Shell executive with a bag of toxic air captured outside her house, which sits 17 feet from the factory fence in Norco."[21]

Richard explained she had once visited Nigeria and that was when she realized the polluted Cancer Alley "is not all that different than the Niger Delta. When I saw environmental destruction in Nigeria, I saw the reflection of my own hometown and cried."[22] Her native soil was the historically African American neighborhood of Old Diamond in Norco, Louisiana, where Richard's house lay so dangerously close to the fence line of a Shell Chemicals plant.

The Climate Justice Conference was just the first of its kind. In 2002, activists met in Bali to hammer out the Principles of Climate Justice. Then came conferences dedicated to promoting climate justice in Bali, Copenhagen, and Bolivia.

By November 2022, the topic of climate justice made its way onto the agenda for discussion at the annual United Nations Climate Change Summit Conference (COP27) held at the Egyptian coastal city of Sharm el-Sheikh, where an estimated 30,000 people from around the world—including diplomats, business leaders, scientists, representatives of NGOs and environmental groups, and others—listened to 100 speakers representing many of those from around the globe most at risk from climate change but who until now did not have a place "at the table."

One of them was now a respected superstar, Dr. Robert Bullard. "In the past, we've come to COP, but we didn't have a place that was ours, a place where we were truly welcome," he told his audience. "Not this year. This year we have a home."[23]

Thus, the work of the justice activists, much of it pioneered in the American South, had been validated by the UN. They now had the attention of the world. But what came next? Was there anything that could really be done about climate justice?

FOUR
IS ECO JUSTICE POSSIBLE?

When activists demand justice of any kind, what are they asking for? In a broad sense, justice, a moral philosopher might reply, means that people should get what they deserve. This is a fine notion until someone wants to know who deserves what? Why? How much?

Such concerns are not new; however, the Australian philosopher Peter Singer says that in preindustrial times the only moral responsibility human beings believed they owed strangers was not to hurt them. After all, until a few centuries ago, most humans knew only a few hundred other beings who were in their clan or village.

Today, though, globalization and the digital revolution have greatly expanded interconnections. The COVID pandemic in 2020 provided a bitter reminder that a huge and dangerous threat demands an integrated and shared response from all nations. But what should that response be? Does each nation have a primary duty to protect its own people first? Or should the international community pool resources and use them to tend to the worst cases first? What moral guidelines should countries follow?

What's Best for Everyone?

Utilitarianism, a popular philosophical notion from nineteenth-century Britain, has long provided many democracies with one way to think about ethical governing. In brief, utilitarianism requires officials to take actions that result in "the greatest good for the greatest number." Using city tax money to build a public park or library for all citizens is utilitarianism.

Using this approach as a basis for climate justice could be fraught with problems. How, for instance, how can anyone decide what is just and necessary for the eight billion people who now live on earth? Or have even the vaguest idea of what the value systems of future generations may be?

Is Equity the Key to Climate Justice?

Concerns like these, therefore, have driven many climate justice theorists to seek guidance from American political philosopher John Rawls, who argued the goal of justice is fairness, not equality. According to Rawls, governments should strive to distribute resources to the least advantaged in society to help them become materially better off than they would have been under a system of strict equality where all individuals receive only their fractional shares, as if they were corporate dividends.

That means policymakers should think in terms of equity, which means fairness, not equality. Thus, the goal of climate justice action for many activists is to make sure everyone receives protection from the ravages of a disturbed climate and has equitable access to help when adverse consequences arrive.

However, there is a problem with this notion. And you get a sense of it every time a political leader proclaims, "When it comes to climate change, we are all in this together, and we all must do our part." To the many millions of people already devastated by a changing climate whose homes are flooded or blown apart by catastrophic tropical storms, or whose crops are crisp and brown in the baking sun, the word "we" rings hollow. In fact, about one billion people on earth go to bed hungry each night. An estimated 3.4 billion others try to survive on slightly more than five dollars a day. Even more startling is that 26 individuals now own as much wealth as the poorest half of humanity.[1]

At this point, the deeper meaning of equity comes into sharp focus. Under guidelines established in the 2015 United Nations Framework Convention on Climate Change (UNFCCC) for a global response to climate change, the richer, developed nations have a greater duty to act because they are responsible for the biggest share of past GHGS emissions. Inarguably, they also benefited the most from the industrial activity responsible for harmful gases. Meanwhile, other nations—often designated as the Least Developed Countries, Small Island Developing States, and countries in Africa—rank among the poorest and most vulnerable and are least responsible for past emissions.

Does this mean the top industrial countries representing 20 percent of the world's populations that are also responsible for around 80 percent of all carbon emissions—and along the way have become immensely wealthy and powerful at the expense of poor countries—have greater burden to fix the climate problem? Do they have climate debt?

Yes, they do, says the United Nations Framework Convention on Climate Change (UNFCCC); moreover, these same nations must reduce emissions and take other steps on a level that is commensurate with how much their past actions led to the climate situation and their current ability to comply with the convention's equity principles. This way of thinking, however, has not settled well with several US presidential administrations. In 2015, negotiators for the Obama administration, for instance, objected to the use of equity as a goal, arguing there was no good method for figuring out what "fair" meant to both developed and developing nations.

As societies everywhere try to navigate a just passage into the age of climate change, new ethical questions emerge. What should be done, for instance, for communities slated to become new sacrifice zones? As some nations move away from fossil fuels, for instance, the demand for minerals such as lithium, nickel, cobalt, and copper—critical to the manufacturing of electric vehicle batteries—is fast growing. Demand, however, is going up for more than just those four well-known minerals. In fact, according to a recent study from the Centre for Social Responsibility in Mining of the Sustainable Minerals Institute at the University of Queensland, Australia, as many as 30 energy transition minerals and metals (ETMS) are needed to make this green transition possible.

A justice problem arises from extracting those minerals, note the study's authors: "Our analysis reveals that more than half of the ETM resource base is located on or near the lands of Indigenous and peasant peoples, two groups whose rights to consultation and free prior informed consent are embedded in United Nations declarations."[2] Other studies show those percentages may be even higher on tribal lands in the United States. The Australian study found these patterns exist in the United States, Canada, Latin America and the Caribbean, Europe, Africa, the Middle East, and the Asian Pacific area. In June 2021, the New York–based finance company MSCI analyzed 5,336 US mining properties with ETM deposits and concluded that most of them are not on Native American reservations. However, the company also noted the majority of the reserves do lie within 35 miles of them. This holds true across the United States, including Texas, Oklahoma, Louisiana, Florida, Virginia, Alabama, Mississippi, Tennessee, and South and North Carolina.[3] Indigenous people often complain that authorities and outsiders fail to consult them or take them seriously, including scientists, when they express doubts about mining operations on their lands, or offer suggestions based on traditional knowledge. Many also have bitter memories of social unrest and violence caused by work crews during past decades of oil drilling. The promise of new jobs and royalties sounds desirable. But who cleans up the mess on tribal lands afterward?

Future Climate Costs

Another concern raised by economists and policymakers is figuring out how much social discounting planners should use in climate change calculations. That is, what share of resources should be available to people alive today, and how much reserved for those of the future? And how do you even devise a method for evaluating the costs and benefits of climate change over time that's fair to both those who are alive now and future generations? What percentage of our Gross Domestic Product should be set aside to pay to protect the yet-to-be-born from future climate woes by planting forests, restoring wetlands, developing new green technology, and building walls to contain sea level rise? How much inflation do you expect to occur in the future? There is just so much money available. If we do provide funds to future generations, how much do we spend this year on them? The year after that?

Anticipating the cost of future climate damage—perhaps in the next

Residents of the predominantly African American community of Warren County, North Carolina, demonstrate against their state government's decision to dump soil contaminated with polychlorinated biphenyls (PCBS) in a local landfill in 1982. Photo courtesy of *Raleigh News & Observer.*

50 years—is yet another worry. Who could have reckoned the damages caused by Hurricane Ian, for instance, with its 140 mile an hour winds and 16-foot storm surge hitting Southwest Florida on September 28, 2022, ravaging Fort Myers, while also damaging Cuba, the Carolinas, and Virginia? According to the NOAA National Centers for Environmental Information, the hurricane inflicted $109.5 billion in damages on Florida alone, making Ian the costliest hurricane in the state's history.

Justice-Seeking in the South

Putting a price tag on damaged property is tough, but it is far harder for economists to estimate social damage such as injury to quality of life, income, and health. There's a lot of thinking about this in the South, where many social justice activists claim there are direct links between the region's racially discriminative past and a wide spectrum of current issues, such as those concerning income, health, voting rights, and environmental injustice.

And there are demands for action. Among the most contentious are those calling for reparations for a legacy of human woes stemming from enslavement and institutionalized racism. The idea isn't new. It first popped up at the end of the American Civil War when Union General William T. Sherman issued a special field order requiring shares of 400,000 acres of land along the Atlantic coast of Georgia, Florida, and South Carolina seized from enslavers by Union forces to be parceled out to the formerly enslaved.[4]

This order was never honored. Others, however, were. Under the Civil Liberties Act of 1988, for example, the federal government paid over 82,000 Japanese Americans $20,000 per person to compensate them for having been forced from their homes, jobs, and schools and relocated to internment camps, following Japan's attack on Pearl Harbor in 1941. Survivors of the camps also received an official apology from Congress.

In 1946, Congress also established the Indian Claims Commission to compensate federally recognized Indigenous tribes to consider claims for loss of tribal lands unlawfully confiscated or other injustices resulting from federal Indian policy. In 1978, the US government paid $1.3 billion to many Native Americans, an amount which provided on average less than $1,000 for each recipient.[5]

For decades, more than 100 tribes were in court, alleging the Department of the Interior and the Department of the Treasury mismanaged monetary assets and natural resources on tribal lands, such as timber and fossil fuel deposits. Finally, in 2016 a settlement of these claims, some more than a century old, cost the federal government an excess of $3.3 billion.[6]

Indigenous people have scored a few more successes. For instance, in 2018 a citizen in Iowa gave private land in Colorado to the Ute tribe. In 2024, California agreed to return state holdings to the Shasta Indian Nation. Meanwhile, authorities in Montana had been negotiating a land swap favorable to the Confederated Salish and Kootenai tribes on the Flathead Reservation. As of 2023, a federal Land Buy-Back Program for Tribal Nations had completed a 10-year process of returning nearly three million acres to tribal trust ownership in 15 states.[7]

Today, many African Americans are still waiting for a similar kind of justice. They haven't been completely ignored. Over the years, in fact, members of Congress have introduced several bills calling for reparations for descendants of enslaved Black Americans. None of these, however, became law. If any ever are enacted, the price tag could be steep. In 2020, Duke University professor William Darity Jr. and his wife Kirsten Mullen coauthored a report for the Roosevelt Institute which laid out a case for requiring the US government to pay up to $12 trillion in reparations to the descendants of those enslaved long ago.[8]

Passing such legislation, in any case, could be the easy part. Next would come the wrenching chore of trying to redress an injustice nobody alive today caused or personally suffered. Whom do you hold responsible for past specific injustices: individuals or institutions? Is it fair to expect a new generation to accept blame, apologize, and make reparations for something they didn't do? And have descendants alive today really suffered substantially from injustices inflicted upon their ancestors? How far back in history can you go? At what point has the debt been paid? That is what the German government wondered in September 2022, when Polish authorities demanded more than $1.003 trillion for the damage done to their country by the German troops during World War II. German officials balked, citing their nation's previous reparations payments to the Poles and an agreement it had made with their government.

Not everyone seeks direct cash payments. Some American activists instead pursue compensation in alternative forms such as college scholarships, student loan forgiveness, and employment initiatives. Across the South also come demands for public apologies, acts of atonement, and the removal of monuments eulogizing Confederate generals. Efforts have also been underway to revise current historical depictions in parks, museums, and government exhibition rooms and to revamp school curriculums to provide a fuller, more accurate account of a troubled past.

Justice of this sort arrived not long ago in Ocala, Florida, after two local prominent Black women, City Council member Mary Sue Rich and Ruth Reed, president of Neighborhood Citizens of Northwest Ocala, led a 16-year-long grassroots battle against a polluting charcoal plant that for 30 years had dirtied the air and spewed soot on the mostly African American community nearby. With help from the nonprofit environmental law organization Wild Law, the citizens' group convinced the Florida Department of Environmental Protection to investigate the Georgia-based Royal Oak's compliance record with environmental laws. In 2006, however, the company suddenly closed the plant. The city council then bought the property and arranged for the destruction of the charcoal plant and the removal of its remains.

The activists were triumphant. There was, however, even more to celebrate on January 10, 2023, when the city government, with great ceremony, opened on the former pollution site a new $10.3 million, 41,750-square-foot, two-story community center that houses a branch library, a banquet hall, two gymnasiums, and a meeting place for seniors. The facility also offers on-site diet and well-being guidance from local medical professionals.

To make sure that people in the years to come will know who the prime movers are behind this effort to transform a toxic hell into a happy community center, local leaders agreed to dub the new facility the Mary Sue Rich Community Center at Reed Place.[9]

Are Reparations Fair and Just?

So far, Southern activists have had some luck at the state level. In 1994, for example, the Florida state legislature for the first time acknowledged that in 1923 state officials failed to respond when a vicious white mob overran the small, rural African American village of Rosewood in the state panhandle, murdering at least six residents and burning houses and churches. To atone, the state legislature authorized $2.2 million in compensation to be divided by elderly survivors of the attack. It has also provided more than 300 college scholarships for descendants of the victims. By providing such payments, the Florida legislature became the first state government in the nation to make financial amends for racial violence, according to Martha Barnett, a Holland & Knight law firm partner who worked gratis on the Rosewood case for the afflicted families.[10]

In 2013, North Carolina lawmakers concluded that their state, too, had a duty to compensate victims of injustice and provided $10 million to survivors—mostly poor, disabled, or Black—of a much earlier state-sponsored eugenics program that included forced sterilization. Three years later, Virginia offered $25,000 to each survivor of a state-sponsored sterilization program of its own.

Reparations for Black Americans at the state level are also being debated outside the South. In fact, it's an ongoing argument in California, which was never a slave state.

But there was an Old South connection in California. Between 1850 and 1890, more than 4,000 enslaved people arrived in the state in the grasp of Southern plantation owners and others who forced them to work in gold mines. In recent years, a California task force has proposed the state pay descendants of these mine workers up to $1.2 million per person as reparation payment for damages resulting from over-policing, mass incarceration, housing discrimination, and healthcare inequity. Unresolved, the issue is still wending its way through the state legislature.

Social activists aren't content to focus all their attention on governments. Commercial corporations with historic links to Southern enslavement are also in their sights. In the past quarter century, in fact, 20 companies within the tobacco, banking, insurance, textile, and railroad industrial sectors faced lawsuits seeking reparations for damages related to alleged enslavement-related practices. One defendant was the insurance giant Aetna Inc., which once wrote life insurance policies for the lives of enslaved people whose beneficiaries were their enslavers. The case, however, failed after it ran into procedural and substantive obstacles. During the court proceedings, the presiding judge raised serious legal questions. Did the plaintiffs have legal standing to sue? Could they prove kinship with descendants of actual enslaved people? Do statutes of limitations apply?

Bryan Stevenson and the Importance of Truth

Truth is arguably the one indispensable thing required for reconciliation. American civil rights lawyer and justice-advocate Bryan Stevenson has long stressed the essential role of veracity in the quest "to repair the damage created by centuries of racial injustice." Doing so, he argues, requires us "to honestly confront our past so that we can understand what remedies are needed to achieve healthy communities and justice for people who have been unfairly excluded and targeted."[11]

In recent years, Stevenson has risen to national prominence as a result of his own social justice work. As founder and executive director of the Equal Justice Initiative, a human rights organization based in Montgomery, Alabama, he successfully led a national effort to provide legal representation to poor people facing criminal charges and those who may have been wrongfully incarcerated. Stevenson was also the prime mover behind the creation of the National Memorial for Peace and Justice in Montgomery, now an unforgettable outdoor, southern landmark featuring a powerful array of rusting steel monoliths hanging from ceilings that bear the names of more than 4,400 African Americans lynched in 12 southern states from 1877 to 1950.

It's best to walk slowly through this disturbing six-acre commemoration to the victims of mass murder, nicknamed "The Lynching Museum." And it doesn't take long to absorb Stevenson's big truth: racial murder is what many white supremacists once stood for. As your eyes drift across all those names of the dead, you may, perhaps, see those of victims executed horribly and unlawfully back in your own home county. Observe, too, the wooden shelves bearing the weight of 800 jars of soil extracted by museum workers from the places where people were murdered by vigilantes and mobs.

"Do you ever have trouble from angry white visitors?" this author asked a docent during his visit to the memorial.

The young Black woman smiled wanly. "Not really," she replied. "But sometimes people want to know where's a display that shows the legacy of white people."

"What did you say to them?"

"I tell him this *is* part of your legacy."

Just a short walk from the memorial stands another Stevenson-led achievement, also dedicated to showcasing terrible truths of the Old South: The Legacy Museum: From Enslavement to Mass Incarceration. The museum merges artwork, scholarship, and innovative technology, such as the use of talking holograms of actors dramatically depicting the injustices of slavery, the horrors of lynchings, and the humiliations of racial segregation. The hope among the sponsors and staff of the museum is that these exhibits can not only educate and enlighten but also create an emotional response in visitors and help them experience the redemptive power of personal contemplation and introspection.

Can Restorative Justice Work?

Not everyone thinks introspection is enough. Some also want accountability. For this, many activists turn to the principles of restorative justice, a victim–offender mediation method pioneered in the 1970s and used worldwide to help perpetrators accept responsibility for the harm they have done to others and make amends. It can either serve as an alternative to criminal and civil law or add another layer of justice-seeking once the courts finish with the matter. Sometimes, restorative justice is practiced outside the courtroom altogether. Human rights commissions, for instance, have relied on it in countries once riven by war crimes and human rights violations that now wish to heal wounds and move forward without triggering renewed episodes of strife.

To initiate the process, a third party sets up a meeting for victims and offenders or representatives of institutions implicated in the conflict. All sides are expected to show good faith during the proceedings, but they may speak frankly. At times, talks conclude with a group consensus regarding what offenders can, or must, do to repair the damage. Apologies, restitutions, reparations, compensation are all possible outcomes.

In a trial, prescribed courtroom rules often prevent ugly truths related to wrongdoings from emerging. During a justice-seeking session, however, victims often have an opportunity to explain in their own words the emotional and physical injury they have endured. When the process works best, offenders acknowledge the harm they caused, take responsibility for their actions, and try to make amends.

Transformative justice—a close cousin of restorative justice—takes the process to another level. Its goal is to get to the truth of the problems that led to conflict in the first place. Then, if possible, steps are taken to eliminate or mitigate the conditions that gave rise to them.

Truth commissions have existed for a half century or so. One of the first was Argentina's 1983 National Commission of the Disappeared that produced the Nunca Más (Never Again) report, documenting human right violations occurring under the military dictatorship in the late 1970s. Over the years, similar efforts have been launched in over 30 countries, including Bosnia-Herzegovina, Canada, Chile, Timor-Leste, Germany, and Uganda.

Tragedy in Greensboro

Southerners have used this approach, too, but only rarely. One of the most painful attempts occurred in the wake of a deadly confrontation on November 3, 1979, in Greensboro, North Carolina. On that day, a group of white and African American demonstrators prepared for a "Death to the Klan" protest march organized by the Communist Workers Party that was scheduled to begin at noon at Morningside Home, a predominantly Black housing project. At 11:20 a.m., a caravan of vehicles appeared, carrying members of the Ku Klux Klan and the American Nazi Party who opened fire with pistols, rifles, and shotguns, killing five protesters and wounding ten others. Television cameras captured grotesque images of the massacre. No police were present.

In 1980, six white supremacists, charged with committing murder at the melee, stood trial in a state court where an all-white jury acquitted them. Four years later, nine defendants appeared in a federal criminal court, all charged with violating the civil rights of slain marchers. An all-white jury exonerated them, too.

A final trial occurred in 1985. This time, a federal civil jury found eight individuals—five Nazis and Klansmen, two Greensboro Police Department officers, and a police informer—liable for the wrongful death of a single victim.

The Greensboro Massacre, as it is still known, remained a festering wound until a group of Greensboro citizens—religious and community leaders along with others—convened and set up the Greensboro Truth and Community Reconciliation Project, modeled on the restorative process used in South Africa. They also invited the community at large to nominate and elect seven members to serve on an independent fact-finding commission, which first met on June 12, 2004.

The Greensboro city government, however, never officially sanctioned the commission, depriving it of subpoena power to summon witnesses. Nonetheless, two years later the commission released a 529-page report that assigned varying degrees of blame to the Klan, the Nazis, the Communist Workers Party, the Greensboro Police Department, the FBI, the Bureau of Alcohol, Tobacco, and Firearms, and the Morningside Homes Community. The report also provided recommendations: the perpetrators of violence should apologize to their victims, and the Greensboro city government should officially acknowledge the incident, erect a public commemorative memorial, establish citizen forums, and host healing workshops. Another recommendation was that the county government should include race relations in its various employee training exercises.[12]

In 2015, the city government erected a marker memorializing the massacre, and two years later officials formally apologized. To many in Greensboro's Black population, such gestures came late and offered little.

Despite widespread disappointment in Greensboro, the idea of restorative justice continued to gain favor across the South. By the fall of 2022, in fact, every southern state had some sort of restorative justice program, notably for schools and the criminal justice systems.

Restorative justice wasn't playing much of a role, however, in reckoning of environmental and climate justice matters. But there was cautious optimism that this was yet possible, because a year earlier the Biden administration had unveiled its Justice40 Initiative mandating federal agencies to "deliver 40 percent of the overall benefits of climate, clean energy, affordable and sustainable housing, clean water, and other investments to underserved communities."[13]

And the administration had an eye on Appalachia, the Gulf Coast, and the coastlines, areas that it considered the most vulnerable and ill-equipped to deal with climate change. To see who was eligible for financial help, federal officials relied on established guidelines as they examined low-income census "frontline" communities with histories of disinvestment and disproportionate environmental, economic, or health burdens. Federal programs loans and grants were then made available for projects involving EV charging networks, freshwater treatment, flood mitigation, community resilience, access to green energy, lead paint removal, and job training programs for clean energy jobs, especially for displaced workers in the fossil fuel industries.

Funding flowed through two dozen federal agencies on its way to state governments instead of going directly to local communities. However, the local level is where many analysts focused their scrutiny. Many were impressed and showered praise on the Biden project for ambition and intention. Nothing had ever been tried like it before. Some wondered, however, how much of the spending was truly addressing the most pressing and fundamental problems in the local communities. How closely were all stakeholders listened to? Were the right kinds of projects getting approval? What good, for example, was an EV charging station if you couldn't afford an expensive EV car? Were there precautions in place to make sure a carbon capture plant wasn't erected in a frontline community? And were these frontline communities, like so many others in Appalachia with its scarcity of resources and infrastructure, able to take on big-scale green projects, even if there was a public will to do so?

Beyond these concerns loomed another far-reaching wall of resistance to the very notion of environmental justice itself, and by 2025 a powerful backlash had flattened many of the achievements of climate justice advocates.

FIVE
PUSH BACK

Hardly had the environmental justice movement got underway before the first signs of backlash appeared. In 1994, Kent Jeffreys, then a Senior Fellow for the Dallas, Texas–based National Center for Policy Analysis, a nonprofit conservative think tank, was one of the first skeptics. The existence of inequities, he argued, didn't necessarily mean they were caused by malicious people. Instead, the availability of cheap land could explain why polluting industries and poor people move to the same place. Industries, he added, tend to locate where they can obtain the cheapest labor. Workers often prefer to live near where they work for any number of reasons. During the Great Migration, for instance, millions of Black Americans migrated to the industrial and polluted North to work, and they lived much better there, Jeffreys argues, than if they had stayed put in the authoritarian, violent, racist, Jim Crow South, even if did have fresher air.[1]

Jeffreys and others also wondered whether the specific cases of environmental injustice cited by activists were all that bad, especially when compared with other hazards that not only minorities, but everyone, must live with. "Much has been said of potentially harmful levels of lead in the bloodstream of inner-city minorities," he wrote. "Yet the primary risk of 'lead poisoning' in urban areas comes from the mouth of a gun rather than a water faucet or a paint can. Murder is the leading cause of death among young male African Americans. While over 400 people were murdered in Washington, DC, last year [1995], not a single person died because of groundwater contamination from a hazardous waste site."[2]

Like Jeffreys, many of the early skeptics were white conservative males, which made them easy targets for some environmental activists eager to cry racism or paternalism. However, there were Black intellectuals who also raised concerns. "It is all too easy to be overwhelmed by clear and tragic historic injustices—and to glide easily from those injustices to a cause-and effect explanation of contemporary issues," wrote Thomas Sowell, a Black economist and social theorist, at the conservative-leaning Hoover Institution.[3] Everyone's life, he argued in his 1999 book, *The Quest for Cosmic Justice,* is impacted by not only race, but also differences in mental acuity, physical attributes, education, child-rearing practices, opportunities, morals, and geography. Trying to "rectify social inequities" with policy would require knowledge that "is staggering and superhuman."[4]

Another African American critic was Christopher H. Foreman, author of *The Promise and Peril of Environmental Justice.* In 2000, the in-house scholar at the left-leaning Brookings Institution argued in the think tank's magazine, "Racial and ethnic disparities are pervasive, and especially distressing in the case of African Americans." Nonetheless, he went on, "when you clear away all the smoke blown over risk and racism in recent years, there turns out to be remarkably little good evidence demonstrating that low-income and minority citizens regularly bear a disproportionate share of society's environmental risk, much less that they develop pollution-related illnesses more often than other citizens."[5]

The Words They Use

Two years later, while testifying before the US Commission on Civil Rights, Foreman critiqued some of the buzzwords used by environmental justice activists, such as fairness, inclusiveness, and even environmental justice itself. They lacked clear definitions, he argued, making them unreliable when trying to craft honest and effective public policy. To Foreman, most of the issues were just another way of saying "not in my back yard."

He did detect an unseen benefit, however. As he also told the commission, "I think environmental enforcement can yield real benefits for low-income communities and communities of color . . . by addressing collective quality of life challenges, guarding not so much against cancer and hypothetical 'endocrine disrupters' but, instead, reducing more prosaic and tangible threats . . . [such as] . . . filth (and the risk of infectious disease that come with it), odors, dust, noise, congestion, the absence of recreational and park facilities . . . [and by setting] . . . three reasonable . . . environmental priorities for communities of color: asthma, childhood lead exposure, and the continuing threat of acute chemical poisoning, faced regularly by some of our nation's poorest citizens, the largely Latino farmworker population."[6]

Foreman wasn't the only one concerned about semantics. Sometimes, the language used by researchers of environmental harm was mystifying, such as a 2019 study published in the *Proceedings of the National Academy of Sciences of the United States of America* in which the authors suggested, "Some may find it intuitive that, on average, Black and Hispanic minorities bear a disproportionate burden from the air pollution caused mainly by non-Hispanic whites." To prove this is so, they provided this exercise in human thought:

> We define and quantify pollution inequity for a group $g(I_g)$ as the fractional difference between a racial-ethnic group's exposure to $PM_{2.5}$ caused by all groups (E_g) and that group's population-adjusted contribution to the overall $PM_{2.5}$ exposure of all groups (C_g) (Eq. 1):[7]
>
> $I_g = (E_g / C_g) - 1$

This definition, along with complex calculations, led the researchers to conclude: "On average, non-Hispanic whites experience a 'pollution advantage': They experience ~17% less air pollution exposure than is caused by their consumption. Blacks and Hispanics on average bear a 'pollution burden' of 56% and 63% excess exposure, respectively, relative to the exposure caused by their consumption."[8]

Few environmental activists talk this way. Instead, they use rhetoric that arouses emotions, sympathy, and ethical concerns. In a 2015 study for the University of Minnesota, researcher and technical writer Chrishandra Taylor found that effective environmental justice speakers were those who

School buses parked in the shadow of a coal-fired power station near the town of Winfield in Putnam County, West Virginia. iStock.com/halbergman.

could appear as credible, trustworthy individuals able to persuade others to support their cause. By doing this, they also hoped to counteract the idea that minorities were uncaring about environmental issues.[9]

Some social environmental justice detractors, however, wonder if concentrating on emotions shades out more pragmatic concerns. Is it always wise, they wonder, for grassroots groups to oppose a proposed new factory in a "throwaway community," especially in an urban area hard hit with unemployment after local manufacturing jobs were sent offshore? "Pushing away good industrial jobs hurts the pocketbook of urban minorities, and, ironically, harms their health in the process," argued David Friedman in a 1998 article for the American Enterprise, a Washington-based conservative think tank.

Friedman also cited research by the University of Pittsburgh physicist Bernard L. Cohen, who concluded that people exposed to hazardous waste and air pollution exposure might lose anywhere from three to 40 days off their total life. Poverty, however, he claimed, took a much worse toll by reducing someone's life expectancy by 10 years or so. "Separating minorities from industrial plants is thus not only bad economics," Friedman asserted, "but bad health and welfare policy as well."[10]

Such reasoning galled many in the environmental justice community. Why, they asked, should anyone have to choose between pollution or poverty in the first place?

Environmental Justice in the Oval Office

Despite the mounting criticism and pushback, activists were guardedly optimistic in 1992 when President George H. W. Bush not only publicly voiced support for their cause but also set up the first-ever Office of Environmental Equity that developed into the Office of Environmental Justice. Two years later, Bush's successor, Bill Clinton, also lent his support by issuing an executive order requiring all federal government agencies to consider environmental justice in their policies.

Even if these presidential actions were merely symbolic "green window dressing," many activists felt encouraged. At last, they thought, government officials at the apex of American power were acknowledging and legitimizing their plea for a kind of justice many Americans had simply not thought about before.

Optimism within the movement, however, waned when the next Bush—George W.—moved into the White House and began chipping away at environmental justice gains. In 2006, the EPA announced it was no longer conducting environmental justice reviews. Nor were federal agencies still required to take into consideration whether negative health and environmental effects impacted certain minority and low-income groups disproportionately. Instead, the Bush administration advanced a different philosophy: environmental justice was not just for minorities, but for "All People."

These actions were early warning signals of a growing backlash-community determined to redefine environmental and social justice matters. In coming years, for example, conservatives would repeatedly argue that government policies should be "race neutral" over other matters such as college admissions and redistricting of voting districts.

Democratic lawmakers, alarmed by G. W. Bush's executive orders, sponsored several bills to counteract them, but none passed. Ironically, they may have even hurt the environmental justice cause. At the time it was politically "a very attractive issue to introduce [environmental justice] legislation around because it threads so many needles: civil rights, environment, social justice, low-income—so many issues," observed Albert Huang, who was then director of environmental justice at the Natural Resources Defense Council. "But for those same reasons, it's a lightning rod for moderates and conservatives because those issues are viewed as the most progressive and liberal of each of those topics."[11]

In 2007, as the second term of the Bush presidency neared its end, the United Church of Christ (UCC) released an update of its toxic waste report from 20 years earlier. These newest findings were mixed. Yes, the environmental justice movement had made progress in recent years, according to the report, but it hadn't been easy for activists to fend "off attacks and proposals that would dismantle or weaken the hard-fought gains made by individuals and groups that put their lives on the frontline." The public now better understood what environmental justice meant, but much work remained. "Polluting industries still follow the path of least resistance . . . For many industries, it is a 'race to the bottom,' where land, labor and lives are cheap. It's about profits and the 'bottom line.'"[12]

The new UCC report also took direct aim at the Bush administration, demanding that "the US EPA end its attempts to roll back environmental justice and take aggressive steps to implement . . . targeted enforcement where the needs are the greatest and where unequal protection place low-income and people of color populations at special risk."[13]

By 2008, the skies seemed sunnier to the social justice community. Bush was gone. His successor, Barack Obama, was a Black American and surely, he would restore what they had lost during the Bush administration. By the end of his time in office eight years later, Obama had indeed earned high marks from environmentalists for addressing climate change.

He had also managed to keep environmental justice alive as a policy consideration. In 2010, for instance, Obama hosted a White House Forum where top government officials mingled with more than 100 activists to discuss environmental justice and how the government could help. The EPA also published various handbooks to guide policymakers with instructions on how to implement justice-seeking as a component of new regulations. Not until the last day of the Obama administration, however, did the agency issue its first and only official finding of actual discrimination—one that was hardly egregious. In this case, the Michigan Department of Environmental Quality stood accused of "'discriminatory treatment of African Americans . . . in the public participation process' related to the permitting of the Genesee Power Station in a primarily low-income, African American neighborhood."[14]

Reversal and Renewal

These modest measures seemed unlikely to be replicated when Donald Trump took the oath of office in January 2017 and put environmental justice on the chopping block. In May 2017, he proposed getting rid of the Environmental Justice Office's entire annual budget for fiscal year 2018. His administration also took steps to relocate the office to a new location, a move that many saw as an overt signal this administration wanted to diminish its importance. Then on September 4, 2020, Trump ordered all federal departments to stop offering training classes focused on diversity, later calling them "divisive, un-American propaganda."

Environmental justice, however, enjoyed a revival after Joe Biden moved into the White House in 2021 and announced a slew of new green, climate change action projects—replete with justice priorities. Some involved the nation's transportation system. For instance, on June 30, 2022, Transportation Secretary Pete Buttigieg stood with local officials at the site of a proposed rapid transit line in Birmingham, Alabama, to explain that the project was the first ever, a one-billion-dollar pilot federal program designed to help Birmingham residents get better access to Interstate 65—a highway that runs through neighborhoods comprising mostly African Americans. A proposed new bus line, for example, if all went according to plan, would re-link neighbors in a community fragmented by the highway and also help many of them get to work and school more safely and efficiently. "Transportation can connect us to jobs, services and loved ones," Buttigieg remarked that day, "but we've also seen countless cases around the country where a piece of infrastructure cuts off a neighborhood or a community because of how it was built."[15]

Indeed, there were plenty of other American cities whose urban communities of color were divided by the construction of 4,000 miles of a federal highway system begun in the 1950s, such as Nashville, Washington, DC, New Orleans, Houston, St. Paul, Detroit, Miami, and Kansas City. According to the US Department of Transportation (DOT), the construction destroyed more than 475,000 households across the country and forced a million people to seek new residences. Everyone inhaled air poisoned by exhaust fumes, small businesses failed, property values sank, and pedestrian landscapes became hellish. The roar of traffic infiltrated every dwelling.

Nobody knows for sure if racism played a paramount or deciding role in selecting these routes. In fact, there were many factors highway planners had to consider when plotting the path of a superhighway at the time, such as the availability of land, real estate costs, and transportation needs. Racism, however, was likely also on the discussion table. Richard Rothstein, a researcher with the NAACP Legal Defense Fund, observes that as early as the 1930s government officials during transportation planning sessions at the federal level brought up the idea of using highway construction for "slum clearance."[16]

The Birmingham project, however, signaled things had changed, at least for the moment. Buttigieg had come to town with a plan to mitigate the existing harm, not add to it. Moreover, it was also part of a larger federal infrastructure undertaking that provided grants for construction projects purposed with something new—dismantling or redesigning of highways for environmental justice purposes. The DOT, for instance, cited a provision of the Civil Rights Act of 1964 to pause construction of a seven-billion-dollar highway widening project of Interstate 45 near Houston in the wake of complaints from local activists who feared the expanded widening of the highway would displace 600 homes, 344 businesses, public and low-income housing, five churches, and two schools, all located in mostly African American and Hispanic neighborhoods.

This was good news for the activists, but it upset proponents of the four-laning project, who complained that delays would prove costly and prevent transportation improvements that millions of motorists and truck drivers needed. The Texas Department of Transportation, in fact, issued a statement claiming the widening effort would "bring much-needed and significant improvements to I-45 to enhance safety and mobility as our state continues to see robust population growth and freight traffic. The project will also replace aging infrastructure and help with flooding issues along the corridor while improving routes that could be used during an evacuation."[17]

Blowback to the Justice40 Initiative

All of this was true, but that wasn't the DOT's current big concern; environmental justice was. Federal efforts in Birmingham and Houston and elsewhere were all aligned with the goals of Biden's Justice40 Initiative.

Biden's initiative also generated opposition, especially among conservatives everywhere. In January 2022 a letter signed by 16 state governors arrived at the White House, scolding Biden: "Excessive consideration of equity, union memberships, or climate as lenses to view suitable projects would be counterproductive. Your administration should not attempt to push a social agenda through hard infrastructure investments and instead should consider economically sound principles that align with state priorities."[18]

The aggrieved governors seemed particularly miffed over the delay in the I-45 project. It was, they complained, "A clear example of federal overreach."[19]

They had no reason to worry. By the time Biden left office in January 2025, the Birmingham road-widening highway project was back on track as planned. In that same year, construction was also robust on I-45, although the Texas DOT two years earlier had committed $30 million to help displaced residents move and find affordable housing.

By now, it was clear that conservatives were attacking environmental justice, not only on its merits, but also because the federal government was behind the idea. Somehow, race also became an issue. In 2022, for instance, during a committee hearing on the Environmental Justice for All Act, several Republican representatives charged that Democrats were using unfounded charges of racism on national issues for political gain. "It's just a dangerous trajectory for us to continue to force this conspiracy of racism on all of these decisions," complained Louisiana's US Representative Garret Graves.[20]

Even television shows for children became controversial. On April 18, 2021, Jamie Yuccas, a host on Nickelodeon's "Nick News," a children's television series, aired an Earth Day special, highlighting several communities with vexing environmental problems. "What do these cities have in common?" asked Yuccas rhetorically during the broadcast. "They're all examples of environmental racism, a form of systemic racism where minority and low-income communities are surrounded by health hazards because they live near sewage, mines, landfills, power stations, major roads."[21]

One of those communities cited in the broadcast was Duplin County, North Carolina, where pig farmers raised two million hogs annually for commercial purposes. "What the hogs do is they use the bathroom, and it's mixed in with a certain water," complained a young Black woman named Fionia, appearing in the documentary, who said she once lived in Duplin Community. "It is sprayed in all the fields in order to grow crops for food. It gets into our pipelines to where our water turns a different color."[22]

While many reviewers praised the aired special for providing educational content to young viewers, the show also angered numerous conservative parents across America who complained that Nickelodeon was trying to indoctrinate young children with environmental justice propaganda.

Justice Activists Attack Environmentalists

By this time, environmental justice activists had not only amassed an array of conservative critics; they had also picked fights with those who should have been natural allies, the nation's prominent environmental organizations. On March 16, 1990, the first big punch was launched by a coalition of communities of people of color, Indigenous peoples, and low-income populations in the American Southwest that the Paonia, Colorado–based *High Country News* magazine described as "The Shot Heard Round the West." It came in the form of a public letter signed by members of a self-described group of "artists, writers, academics, students, activists, representatives of churches, unions, and community organizations" affiliated with the Albuquerque-based, Southwest Organizing Project (SWOP). Many of the signatories were Southerners who represented kindred organizations in Texas, Virginia, Georgia, Alabama, and North Carolina.[23]

The SWOP message also appeared in the mailboxes of the nation's major conservation groups, at the time commonly called the Big Ten: the Sierra Club, the Sierra Legal Defense Fund, National Audubon Society, the National Wildlife Federation, Environmental Defense Fund, Environmental Policy Institute, Friends of the Earth, Izaak Walton League, the Wilderness Society, the National Parks Conservation Association, and the Natural Resources Defense Council. In clear, accusatory language, SWOP complained that they—the region's people of color—had suffered a legacy of "racist and genocidal practices including the theft of lands and water, the murder of innocent people and the degradation of our environment."[24]

The coalition criticized not only the mining companies for depleting resources and polluting the environment and the US military for conducting harmful operations in Indigenous communities, it also told off the Big Ten, who "often claim to represent our interests, [but] in observing your activities it has become clear to us that your organizations play an equal role in the disruption of our communities."[25]

Until now, environmentalists generally thought they'd done something good by persuading the federal government to set aside certain lands in the Southwest as a national monument. Wrong, argued SWOP: Pueblo people of Acom considered that territory sacred. SWOP also criticized environmentalists for lobbying the federal government to prohibit sheep grazing in Southwest wilderness areas to prevent harm to the natural ecology. This ban, claimed the justice activists, would hurt local Chicano sheepherders economically. They also criticized the environmental organizations for receiving financial support from corporations such as British Petroleum and Exxon, all "known polluters."

Then came another big slam. The Big Ten, said SWOP, were guilty of not including "people of color in decision-making positions" within their organizations. "Racism is a root cause of your inaction around addressing environmental problems in our communities." SWOP demanded immediate changes in how the Big Ten conducted their business: cease operations in communities of color until "you have hired leaders from those communities to the extent that they make up between 35–40 percent of your entire staff."[26]

Across the nation, mainstream environmentalists reacted to the broadside with shock and anger. "I don't think it's possible to overestimate the [letter's] impact," Alan Spears, a National Parks Conservation Association (NPCA) legislative representative, later told *High Country News.* "Very few people at NPCA had any real sense of the environmental justice movement and how it was evolving, and they were very surprised that anyone felt NPCA was a problem."[27]

Internal debates at the conference tables of the Big Ten soon got into full swing. Was SWOP right? Or was its attack unmerited and unfair? Either way, what was the best way to respond to the letter?

Instead of mounting a defense against the charges leveled against them, the Big Ten, along with other organizations both private and public, for the most part, got busy implementing new racial equity statements, policies, priorities, and hiring practices to satisfy their critics. These revisions also appeared across the nation at Sierra Club local chapters, including those in Florida, Georgia, and Virginia that pledged to pursue various justice-centered environmental goals.

Seeing the Big Ten in distress, meanwhile, delighted conservative commentators across the media landscape. "Even avowed socialists have noticed the elitist nature of traditional environmental histories," argued the critic, Kent Jeffreys. "The EPA has typically responded to an elite constituency, not minorities or the poor," he said. "In most cases, the agenda of the environmental elitists does not coincide with the interests of minorities."[28]

Over the decades, other environmental justice groups have piled on top of mainstream environmentalists with their own litany of concerns. In June of 2020, for instance, a Puerto Rican attorney, Elizabeth Yeampierre, the first Latina Chair of the US EPA National Environmental Justice Advisory Council, told the Yale School of the Environment that the country's top environmental groups "were built by people who cared about conservation, who cared about wildlife, who cared about trees and open space and wanted those privileges while also living in the city, but didn't care about Black people. There is a long history of racism in those movements."[29]

That long history, accused critics, includes past prominent historic figures who were alleged racists, such as the National Audubon Society's namesake, James Audubon, an enslaver, and founder of the Sierra Club, John Muir, whose writings contained derogatory comments about Indigenous people and Black Americans. Other luminaries received condemnation for purportedly having kept company with racists or promoters of the now discredited eugenics movement of a century ago. Some justice activists also complained that white people filled too many jobs in the top positions of environmental organizations, thereby making them non-inclusive and thus racist.

In recent years new complaints have come from *intersectional* environmentalists, mainly Gen-Zers (people born 1997 to 2012) who demanded, among other things, that mainline environmental organizations immediately take extra reforms to counter what they asserted was the cumulative toll caused by discrimination and oppression in multiple overlapping ways on marginalized employees, that is, people of color, including those with physical disabilities or members of the LGBTQ+ community.

All this criticism prompted varying degrees of in-house reform in some big environmental groups. In 2020, for instance, two top officials of the National Audubon Society resigned after an in-house survey revealed most employee respondents "felt Audubon doesn't create an environment where diverse staff can thrive."[30]

The strife, however, also inflicted injury. Many of these same organizations saw a huge drop in membership, prestige, and funding. According to the *New York Times,* the Sierra Club, for instance, has lost 60 percent of the 4 million members it had in 2019 and also struggles with an estimated debt of $40 million.

Critiquing the Critique

So, was all this criticism valid? Were the nation's top environmental organizations truly guilty of racism? And did it follow that comments made by conservationists more than a century ago had forever tainted the culture and ethics of the organizations they fought hard to create? Or that today's white, unpaid volunteers in environmental groups, many of whom had also participated in civil rights crusades and fought for various health, economic, education, and childcare reform campaigns to advance the lives of others, were racists?

Being accused of racism then pushed aside to make way for newcomers with a different agenda doesn't usually sit well with the accused. Says one white male executive director of a major Florida environmental group with ties to the major environmental organizations who prefers anonymity, "Tension in the Big Ten over environmental justice is a big issue. It dominates the workplace. Everyone is scared." In fact, some resentful older environmentalists claim a new kind of redlining is being used to disqualify white males from positions of leadership in the Big Ten that's based on sex and race. The environmental justice movement, complains the Florida executive director, is "co-opting the mainline environmental movement." Feuds over allegations of diversity, equity, inclusion, and justice, in fact, have roiled the membership and leadership of many national environmental nonprofits such as Defenders of Wildlife and the Natural Resources Defense Council.

At times, there's also friction between organizations. It's hard to miss the ire in the voice of the activist for a multi-southern state environmental justice organization, who identifies himself as a Hispanic minority, when he recalls how a major environmental organization dispatched one of its own to personally check out the southern justice group's request for financial help. "So, they send some white guy in from Nevada down here into the South to investigate the environmental justice problems we have, and right away it's clear he doesn't know anything about the region or what's going on in the front lines. We're the ones that do!"

All the same, the question remains: isn't there an ethical responsibility to monitor how the money donated by others is spent?

Sometimes, tensions snapped in public, as they did on July 22, 2020, after Michael Brune, executive director of the national Sierra Club, penned a "nostra culpa" [we're guilty] on behalf of the entire organization which he posted on the club's website. "It's time to take down some of our own monuments," he wrote, "starting with some truth-telling about the Sierra Club's early history." The club had, he opined, played a flawed part in perpetuating white supremacy and trying to preserve wild places while forgetting some were ancestral homelands to Indigenous peoples.[31]

To fix things, Brune announced, "[We] are redesigning our leadership structure so that Black, Indigenous, and other leaders of color at the Sierra Club make up the majority of the team making top-level organizational decisions. We will initiate similar changes to elevate the voices and experiences of staff of color across the organization." Brune also pledged to seek the use of $5 million from Sierra's budget "to make long-overdue investments in our staff of color and our environmental and racial justice work."[32]

Many Sierra Club members praised Brune's plan, such as someone who signed in as Ruth Hung Cooperrider on the club's website: "As an Asian American, I have personally felt the pain of racist attitudes and behaviors. I agree it is time for reckoning and reconciliation by the Sierra Club, and I fully support the future steps that your statement promised."[33]

Other posts expressed fury, claiming the Sierra Club was virtue-signaling. "[Brune's] article and its attempt to be politically correct in the current environment has reached a new low," complained respondent Richard Briscoe. "I won't plan to renew my membership with Sierra Club."[34]

The turmoil in the Sierra Club was still underway in November 2022 when Ben Jealous, a prominent civil rights attorney, became the first person of color to serve as the executive director of the stressed organization. From the onset, he faced a slew of internal problems such as demands from employees for pay equity and greater inclusivity, along with allegations of misconduct, and complaints of unfair labor practices from the bargaining units of the club's organized workers. By May of the following year, the club also faced a $4 million shortfall; there was no choice but to lay off dozens of employees, including Black workers.

A Southern Black Woman Stands in Two Worlds

Virginia Sanders was one of them. Before the layoff, Sanders, a native South Carolinian, seemed to have all the credentials needed to be the kind of employee the Sierra Club needed as it restructured itself. After all, she was an African American and a veteran of multiple environmental battles in her home state, including a major effort to stop construction of a proposed 4,600-acre "city-within-a-city" south of Columbia, South Carolina, in a floodway. Developers said the project would be surrounded by state-of-the-art levies which Sanders and others feared would cause flooding in nearby mostly Black communities. She had also tackled several other regional water issues related to surface and groundwater pollution, and runoff—all of which adversely affected people of color. "We're tired of being treated as collateral damage," she says.[35]

Roaming the world as an army wife, observing various peoples and cultures, and comparing American behavior toward nature with the way it exists in other lands gave her perspective. Best of all, she also had environmental justice experience. "I was a child of the civil rights movement," she points out, "and I've long been an active member of the of Lower Richmond NAACP."[36]

Such a résumé made her an ideal choice as a National Sierra Club Organizer to help advance environmental justice in the Richland and Marion Counties of South Carolina. That, however, didn't protect her from the layoff.

She did see it coming, though. "I wasn't blindsided," she says. "I knew something was happening, so I was prepared. Sierra used to be operated from the bottom up. I liked the grassroot flavor in it. But beginning about four years ago, I heard rumblings that we can't operate that way anymore. I understand the complaints of the young people in Sierra, but I can't say that I have experienced any of those things myself. For many years I was the only Black person in the Sierra Club in South Carolina, and during that time I attended conferences all over the state and nobody ever said anything bad to me or treated me badly."[37]

Perhaps, she now wonders, the upheaval in the Sierra Club represented a tension that inevitably happens whenever the next generation shows up to challenge the status quo. "In some cases, older people should be stepping aside and passing the torch, but the younger people need to know what to do with the torch."[38]

Today, Sanders is still busy, working on environmental issues and helping to prepare the next generation of environmental warriors to tackle recent problems related to the destruction of natural habitat and pollution of the state's freshwater supplies stemming from new developments in the fast-urbanizing regions along the I-95 corridor running from Charleston, South Carolina, to Jacksonville, Florida. Not long ago, she spoke with officials at two "historic universities" in Columbia, the state capitol, where she shared her experiences with students. "I want to download my brain. They need the institutional knowledge."[39]

The Shakeups Continue

Another blow, meanwhile, has struck the Sierra Club. It came on April 16, 2025, when Robert Bullard announced that after working with the Sierra Club since the mid-1990s he was calling it quits. He acknowledged the federal government had negotiated a voluntary agreement with the Alabama Department of Transportation (ALDOT) requiring the department to fix "the flawed four-lane US 84 flooded the . . . [Black Shiloh community in Alabama] . . . and left the residents and their homes to drown. . . . [However, he complained the feds hadn't made good on a promise to make ALDOT compensate] property owners for six-plus years of highway flood damage, leaving residents' severely damaged homes to sink into the Alabama mud."[40]

Adding insult to injury, in Bullard's eyes, was that Sierra Club Executive Director Ben Jealous "stood on land in the Shiloh community, land that's been in some families since Reconstruction, and promised to secure an attorney for the homeowners and support a lawsuit against ALDOT seeking compensation for damage to their property—empty promises that never materialized."[41]

In August of that year, the Sierra Club's board of directors terminated its contract with Jealous. As of early 2026, the club was reportedly still in turmoil.

And now decades after the debut of SWOP's divisive open letter, many environmentalists and justice activists are left wondering if all the fighting, posturing, and accusing is accomplishing much, especially with climate change unfolding. Writing in the *Atlantic* magazine, Jedediah Britton-Purdy, a law professor at Columbia, suggested that if everyone calmed down and took a "long" historic view they would find that mainline environmentalists were not interested in waging racist crusades. Instead, he writes, "For more than a century, activists and scholars have been engaging in the themes of fairness, inequality, and political and economic power in the human environment."[42]

Even if the accusations coming from the environmental justice community have been overblown, few could deny that for years the human factor has often been overshadowed in courtrooms and administrative hearing rooms, where mostly white attorneys representing environmental groups, government agencies, and corporations typically debate interpretations of the minutiae of federal and state laws, policies, and administrative rules set by regulatory agencies focusing on environmental science. Anyone wanting to insert justice concerns into most of these proceedings was simply at the wrong forum.

One thing is clear, however: the environmental justice movement shows no sign of fading. In fact, observes author and climate change planner Jim Schwab, "The new movement had won a place at the table. The Deep South, the nation, would never discuss environmental issues in the same way again."[43]

If so, justice activists and mainstream environmentalists are likely to recognize the need to form new alliances. Pat Bryant, executive director of the Gulf Coast Tenants Organization, recognized this at the First National People of Color Environmental Leadership Summit in Washington, DC, in 1991. At the time, he said, "I think there is fertile ground for coalition and cooperation. But it cannot happen unless we adhere to some very basic principles. . . . We cannot join hands with anybody who will not join with us and say that we have the right to live. . . . We need our friends who are environmentalists to look at a total program for human uplift."[44]

Today, however, some younger activists signal a reluctance to reach consensus with traditional environmentalists. "They can stand beside us, if they want to," Leah Thomas, a Missouri-based founder of the Intersectionality Movement—a campaign that stresses the "inextricable link between environmentalism, racism, and privilege"—told *National Geographic:* "If you're not going to offer me a seat at the table, I'm going to create my own, and hopefully we can have more and more organizations cropping up. We've grown a community of over 70,000 young people, and it's just growing and growing. They can find a way to partner with us."[45]

Making ultimatums like this, however, may prove counterproductive. Both camps need all the help they can get.

Meanwhile, there are legions of people who pray the disagreement continues because, in their view, anything that thwarts environmental and climate justice action is good. That's because they insist, all the rhetoric about environmental and justice problems is overblown and climate change itself is nothing but a hoax.

SIX
ROOTS OF DENIAL

On September 23, 2019, people from around the world gathered in New York City to attend the latest UN-sponsored climate summit. They came to achieve a common goal—preventing the mean global temperature from rising by more than 2.7°F above preindustrial levels. Outside the air-conditioned conference rooms temperatures soared to record levels.[1]

Meanwhile, about 750 miles away to the south, a far different kind of conference was underway in Louisville, Kentucky. At this gala affair more than 100 lawmakers, state and federal government officials, along with representatives of the oil, gas, and coal industries congregated at the fifty-ninth annual meeting of the Southern States Energy Board. Nobody here was urging to cut carbon emissions. Instead, guest speakers praised the economic role of the fossil fuel industry and called for deregulation of the industry. One of them implored everyone present to do what they could to protect the world's "poor from harmful climate change policies."[2]

By this time, such rhetoric was hardly rare. In fact, it was fast becoming a tenet of faith among conservative Americans, especially for Republicans. Dismiss, delay, denigrate, downplay—all were in the denialist toolkit. Consider the wordplay of Republican Marco Rubio who, while a Florida senator, declared during a 2016 televised Republican presidential primary debate: "Sure, the climate is changing. And one of the reasons [is that] . . . the climate has always been changing. There has never been a time when the climate has not changed."[3]

Since the Greek philosopher Heraclitus had already figured out that *everything* is always in a state of flux back in the sixth century BCE, Rubio's observation may not have impressed many in the audience as being all that revelatory. What he said was essentially correct. Yet it was also beside the point. What concerns climatologists today is *how fast* the change in climate is occurring and *why*. Though their estimates vary, they agree the speed of global warming today is at least 10 times faster than at the end of an ice age that peaked 21,000 years ago—and humans are to blame.[4]

A Georgia Tech Maverick

Most climate skeptics are not professional climatologists. However, some are and they aren't all cranks, insists Judith Curry, Professor Emeritus and former chair of the School of Earth and Atmospheric Sciences at the Georgia Institute of Technology. Nor is their skepticism, like hers, ideological or rooted in tribalism. Instead, says Curry, they are professionals who are doing their job, which is to question the research and conclusions of their peers.

Curry sees herself as one of those climate skeptics—and one of the most outspoken and controversial. But that was not always the case. In fact, she is a noted climatologist who once authored reports for the IPCC. While she agrees in her many interviews, podcasts, and talks that anthropogenic warming is occurring, she is also wary of, as she puts it, the "manufactured consensus of scientists at the request of policymakers." The climate models are not dependable, she asserts; moreover, Curry is dubious about the elevated level of certainty her colleagues express when they set dates for a coming catastrophe. This overconfidence, she believes, has led to widespread groupthink in both the scientific and policymaking corridors. Climate modeling, Curry insists, relies on statistical analysis based on variables that are easy to tweak when setting up the computer model; often, there is not enough attention focused on "outlying variables such as water vapor, melting ice sheets and natural causes such as the impact of sun, volcanoes and slow circulations in the ocean." In short, she says, the world is "too complex" to be captured accurately by human modeling.[5]

Most of her peers disagree.

Outside of the academy, millions of other deniers reject, often with hostility, the findings of the IPCC and other prestigious scientific organizations such as the EPA, NASA, the US National Academy of Sciences, and the American Association for the Advancement of Science. Unlike Curry's professional skepticism, theirs is all too often rooted in pseudo-science, religion, cultural norms, political tribalism, and wishful thinking along with a bounty of propaganda from the fossil fuel industry.

Many who embrace all this live in the American South, a region where denialism has long been an accepted way to think.

Down Home Skepticism

Climate skepticism, in fact, may be an offshoot of an older and far-reaching distrust of environmental activism. And this is puzzling. After all, as many Southerners themselves are quick to point out, they love and respect the natural world, especially where they live. Such fondness has a long history.

"For nearly three hundred years before the American Revolution, the colonial South was a kaleidoscope of different people and cultures," writes history professor Timothy Silver of Appalachian State University. "Yet all residents of the region shared two important traits. First, they lived and worked in a natural environment unlike any other in the American colonies. Second, like humans everywhere, their presence on the landscape had profound implications for the natural world."[6] The South, according to the US Census Bureau 2010 Rural America Study, had the highest total rural population of all regions, with 64.4 percent of the nation's rural population living east of the Mississippi River. In recent years, these figures have dropped a few percentage points because of urbanization and changes in how the bureau defines an urban area. But even many urban Southerners still nurture their rural roots, which may influence their perception issues concerning the natural world. In fact, there's a substantial case to be made that your proximity to nature could play a role in shaping your environmental views, including what you think about climate change. "Voters from rural America are more likely to view where they live as being an important part of how they define themselves which in turn shapes their views, including on environmental policy," according to the authors of a 2019 Duke University study. "Attitudes about the government are a clear dividing line between rural and urban/suburban voters, particularly regarding differences in the level of trust toward the federal government."[7] This same partisan divide also seemed to exist over perceptions of extreme weather. "Further," wrote the Duke University researchers, "we found some rural leaders reticent to discuss climate change within their communities due to the polarization of the issue."[8]

This didn't mean they were hiding their heads in the sand. In fact, rural skeptics were found to be "relatively sophisticated" about environmental problems and acknowledged the need for conservation and regulation. However, they were warier of government policies, especially at the federal level, than their counterparts in cities and the suburbs. And they had their reasons. Many of the rural North Carolinians interviewed by the Duke researchers (with help from the Environmental Defense fund) resented "one size fits all" governmental environmental regulations.[9] What might work in rural areas far away could be problematic in their communities, they pointed out. There were also complaints that policymakers didn't consult with or listen to farmers and others who were expected to abide by regulations on chemicals used on their fields. "Many rural Americans from interviews and focus groups worried that climate policy will leave them out of the conversation and increase hardships in rural communities," observed the researchers.[10]

Faith over Fact

Being informed, as many of those interviewed revealed, didn't necessarily mean having an overriding trust in science. Religion in the South—the Bible Belt of America—however, is a mighty molder of perspectives, secular and supernatural. Clay Higgins, for instance, a conservative Republican congressman from Louisiana, spoke for many faith-based Southerners at a congressional hearing on climate change when he made the case for denialism by saying "a biblical perspective" proves God gave man dominion over the earth. Others, however, may wonder that even if Higgins is right, how does it logically follow that human misuse of that authority is a wise course of action?

To view perplexing issues through the lens of religion—one steeped in conservative, Protestant Christianity—as Higgins apparently does, is common in much of the South. Today, the region is the nation's most religious, according to a 2016 Pew Research Center study. Fourteen out of the top 15 states on the Pew list are southern, with each registering at least 61 percent of its adult population as "highly religious."[11]

Many church-going Southerners who reject Darwinism, for example, do understand the basics of the theory of evolution and may even squirm when their pastors demonize the great biologist. Being related to other primates isn't what really bothers many of them: instead, it's the idea that if Darwin is right, human life is just one possible arrangement of atoms, a result of random events that began a billion years ago. We're only here because a fireball landed and took out the dinosaurs. If all that were true, there was no necessary reason for anyone to be alive. No divined purpose for being here.

That's scary news. Christianity, at least, provides solace. According to Christian assertion, if you say and do the right things, God will bless and reward at least your soul. Thus, a good insurance policy is to accept what the preacher says when science apparently contradicts religious belief.

Some of today's thumb-nosing at science may be traceable to the 1925 Scopes "Monkey Trial" in Dayton, Tennessee, which "stemmed from the modernism rising in the post–World War I era," writes Laura Ellyn Smith, a history professor at Arizona State university. "Southern whites felt that these changes challenged their way of life, including seeing the teaching of evolution as an attack on traditional values. They moved aggressively to retain socio-cultural control in a time of transformative change by limiting modern influences."[12]

Although the modern South is home to every faith in the world, its most powerful spiritual forces remain fundamentalism, Pentecostalism, and above all, evangelicalism—a movement based on a Protestant Christian doctrine shared by various denominations whose adherents believe the Bible is to be understood literally.

Morality instruction for many Southerners, too, comes from scripture, not legislation, or science-based findings. "In modern culture, the evangelical movement has been the most powerful carrier of this kind of religious anti-intellectualism and its antinomian impulse [a doctrine asserting faith is necessary for salvation] and trumps any moral law," observed Richard Hofstadter during the early 1960s in his Pulitzer Prize–winning volume, *Anti-intellectualism in American Life.*[13] To drive home his point, Hofstadter cited the words of North Carolina's Billy Graham, America's great apostle of evangelicalism at the time, who regularly decried the sins of modernism to arena-packed audiences. "[In place of the Bible]," Graham thundered, "we substituted reason, rationalism, mind, culture, science worship, the working power of government, Freudianism, naturalism, humanism, behaviorism, positivism, materialism and idealism."[14]

Many modern Southerners consider warnings from the IPCC about climate change as just the latest utterances of scientific paganism. And even if they see signs that climate change is not a fiction, many pastors encourage their flocks not to worry. If God wants climate change, a fervent fundamentalist Christian once told this writer, that's what He will have.

Katherine Hayhoe, an evangelical Christian and an atmospheric scientist, thinks what influences supposed faith-based denialists is more ideological than theological. As she told *Rolling Stone* magazine in March 2020, "Their real objection, 99.9 percent of the time, is that they don't think there are any solutions to climate change that are consistent and compatible with their values and their ideology." She adds, the "elite cues" for many churchgoers are not coming from the pulpit but rather sources such as "Fox News commentators and . . . the social media pundits."[15]

Admitting the reality of climate change, she also points out, obliges people to accept responsibility for doing something about the problem. Not doing so may make them feel guilty. So, they deny the problem.

Hayhoe is a woman of impressive academic achievements. Among other things she is the Chief Scientist for The Nature Conservancy, a Paul Whitfield Horn Distinguished Professor and the Political Science Endowed Chair in Public Policy and Public Law in the Department of Political Science at Texas Tech University, and a contributing author of the three National Climate Assessments.

Hayhoe is also the wife of a pastor in Lubbock, Texas, and has spent years speaking at forums big and small trying to persuasively but gently convince her spiritual kin to reconsider their denialism. It hasn't been easy.[16]

Habitual Denialism

Nurturing the kind of denialism Hayhoe confronts may have roots in a widespread mental habit baked into Southern culture long ago. White Southerners have pushed back against criticism from the outside world since antebellum days. A useful shield was denialism. "'The peculiar institution' [enslavement] depended on a peculiar kind of denial," writes Colin Dickey, author of *Under the Eye of Power: How Fear of Secret Societies Shapes American Democracy,* "one in which the grossly unnatural condition of bondage had to be misrecognized as a benign divinely sanctioned institution."[17]

This denialism lived on after the Civil War and still does. Known as the Lost Cause Myth, it has provided aggrieved Southerners with a way to sanitize the horrors of slavery and fratricidal war, legitimize insurrection, romanticize plantation life, and sentimentalize the loss of an idyllic, storybook civilization that never truly existed.

A takeaway from research conducted by Kristin Lunz Trujillo, assistant professor at the University of South Carolina, may partly explain the endurance of such thinking. Rural residents, she concludes, may see themselves as an outgroup that contrasts itself from anyone they believe looks down on them—such as urban officials and elites.

In October 2024, a totally new twist in climate denial ideology swept through the digital universe in the aftermath of Hurricane Helene, which made landfall as a Category 4 storm and killed at least 230 people and inflicted possibly tens of billions of dollars in damages. Call it, perhaps, denial by scapegoating others with false claims. This seemed to be what far-right Georgia congresswoman Marjorie Taylor Greene was up to on the social network platform X when she implied the US government had extraordinary and nefarious meteorological power in October 2024. "Yes, they can control the weather," she posted. "It's ridiculous for anyone to lie and say it can't."[18] Worse yet came the wild implications of internet "influ-

A submerged van in the Swannanoa River near Biltmore Village in Asheville, North Carolina, after Hurricane Helene hit the state in late September 2024. iStock.com/FS.

encer" Matt Wallace, who also posted on X a map showing the destructive path of Hurricane Helene across Florida, Georgia, the Carolinas, Tennessee, Kentucky, and Virginia—with text insinuating that somehow the storm had been directed "to almost methodically miss the bluest parts of those crucial swing states, while simultaneously ravaging the red parts," to disrupt fast approaching national elections.[19]

It is true that climate scientists have conducted small-scale experiments such as seed clouding to develop ways to mitigate the harm of climate change; however, no proof yet exists that any government, so far, has amassed the power to weaponize weather for political gain.

Then-candidate Donald Trump alleged at the time that the federal government was confiscating the properties of residents hurt by the storm. In the wake of this misinformation, FEMA had to suspend relief efforts in hard hit western North Carolina because of threats from armed militias.

In a televised address, Biden angrily responded to Trump's words, calling them "reckless, irresponsible and relentless promotion of disinformation and outright lies that are disturbing people. It's undermining confidence in the incredible rescue and recovery work that has already been taken and will continue to be taken. It's harmful to those who need help the most."[20]

(Two weeks later, on October 9, 2024, Florida, still reeling from Helene's assault, received another devastating blow from Milton, a dangerous Category 3 hurricane. Both storms were fueled by ultra-high 88+°F sea surface temperatures.)

Guilt-Inspired Denialism

Trauma also plays a role in Southern denialism, suggests Richard Grant, an English author who took a crack at understanding Southerners by living for a while in Natchez, Mississippi. Southerners, he concluded, continue to suffer the "psychological pressures of Southern history." Black Southerners, he argues in his 2020 book, *The Deepest South of All*, experience these pressures the most powerfully, but their white neighbors "are required to reconcile their history of enslavement, raping, and lynching with their proudly held beliefs in honor, graciousness, and Christian decency."[21]

Grant, in fact, thinks life is a bit tough for many white Southerners. How can they ever explain away the idiocy of succession and the devastating destruction of war that came with it? And how do they go about squaring the many admirable qualities of being a Southerner with the ugly ones? It might be easier just to not even try. "White Southerners have denied the facts and created self-serving mythologies," writes Grant. "They have blamed outsiders and blamed 'the Negro.' They have hurled themselves into religion, craving innocence and absolution, and into the whisky bottle, craving release and oblivion . . . [all of which have had a] warping effect on reality."[22]

Don't forget miscegenation, Grant adds, which is widespread in the South. Race-mixing, however, he notes, was taboo, and often adulterous, "so white society, and white women in particular, performed mental gymnastics to pretend that the abomination wasn't happening, and certainly not on the scale that it actually was happening, even though the living, breathing evidence was all around them."[23]

In 2008, racial animus also may have fueled denialism. Evidence for this notion emerged shortly after democrat Barack Obama—a mixed-race person that die-hard racists so feared—won the presidency. That, at least, is the conclusion in a study appearing in the journal *Environmental Politics*, published a decade afterward. A turning point, argues political scientist Salil Benegal, occurred when Obama called for climate action during a State of the Union address. That's when many conservatives, says Benegal, became more strident in their denialism. In September 2016, their views intensified when Obama later brought the United States back into the Paris Climate Change Accord. According to Benegal's analysis of the election, based on surveys on racial resentment and attitudes about climate change, deniers were more likely to be older, white, Republican, and those who harbor racist beliefs.[24]

Benegal's work is part of a growing body of research that suggests many Americans oppose climate action because they see it as yet another example of unfair government assistance to people of color at home and abroad. The race of the person who is making a point about climate change may influence the mind of the listener, as well. If Donald Trump, for instance, tells racialized white supporters that climate change is a hoax and any money spent fighting it is a waste of their money, they tend to believe him. However, say researchers, American Black people are more likely to embrace a call for climate action if they are listening to a man they trust, such as Barack Obama.

All these insights matter, argues Eric Parajon, a political scientist at the University of North Carolina, specializing in American public opinion on climate policy, "because politicians on both sides of the political aisle can choose to frame climate policy as racial and lead to an activation of racial resentment." Moreover, a better understanding of the vast variety of racial attitudes and prejudices, he adds, provides vital insights for policymakers struggling to "shape efforts to increase climate concern and support for climate change mitigation policies that benefit all Americans including communities of color."[25]

Denialism of the Worst Kinds

As scholars try to understand denialism by analyzing the way people think, there's another kind that calls out for scrutiny, too. It comes from *not thinking*, and Dr. Gary Zarillo, professor of oceanography and ocean engineering at Florida Institute of Technology in Melbourne, says it's "the worst kind of denialism." Zarillo encounters it almost every day along the Florida coast, where he does much of his research. It is powered by incurious or indifferent politicians and people in the world of commerce who continue to promote more development in the form of condominiums, hotels, and shopping centers often erected along the ocean's edge. "It actually gives me pain," he laments. "The new construction are hurricane targets, future sea level rise victims."[26]

Synthetic Skepticism

A final ingredient of the climate denial recipe is, of course, *manufactured doubt* which results from a massive, purposeful, and well-financed propaganda campaign to sow the seeds of doubt, perplexity, and confusion over the science-backed case for climate change. It is funded by fossil fuel companies such as Exxon Mobil and BP and denial groups like the Heartland Institute, the Cato Institute, and the Competitive Enterprise Institute. For decades, the powerful individuals who run these institutions have known climate change was threatening the natural world and that a new "green" economy based on renewable, carbon-free energy systems was emerging in response. They also understood their industries stood to lose enormous profits from lost revenues and the value of $100 trillion worth of assets worldwide that could be "carbon stranded" if the world ever transitioned to a cleaner industrial system. Such assets include fossil fuels still in the ground, along with a vast system of infrastructures, including power plants, refineries, pipelines, ocean platforms, mining equipment, conveyors, trucks, trams, and more.

Thus, some energy titans decided years ago that to protect their corporate and individual interests they had to use their immense influence not only to delay or derail regulation of their industries, but also to misinform hundreds of millions of American about the true and growing danger of climate change.

An even more aggressive and ruthless kind of denialism has taken the stage in recent years: the *outlawing* of climate actions of others. For help in putting this and all other necessary forms of denialism into play, the fossil fuel industries have turned to the Republican Party.

SEVEN
THE GOP VS. CLIMATE ACTION

There was once a time when many Americans shed partisanship and agreed that the planet was in peril and needed protection. That took place on April 22, 1970, when an estimated 20 million people—or about 10 percent of the US population at the time—from all walks of life gathered in cities and towns from coast to coast to participate in rallies, teach-ins, and cleanup events to raise awareness about the country's pollution and environmental degradation problems.

It was the first Earth Day.

And by all accounts it was a smash success. For one thing, all that energy and action inspired politicians, both liberal and conservative, to tackle the nation's environmental problems with new laws. Earth Day also sparked the environmental movement. Today it is celebrated in more than 192 countries and is the largest secular observance on the planet. Earth Day 2016 was chosen as the day when over 120 nations signed the Paris Agreement, an international treaty on climate change.

This harmony in the United States, however, didn't last. As liberals strove to enact ever more green legislation at state and federal levels, many conservatives turned away, complaining about meddlesome activist "tree-huggers" whom they suspected of exaggerating environmental risks and ignoring the financial costs to businesses and communities hurt by regulations. Some doubted regulation was the best policy to fix environmental problems. Maybe taxes on pollution would work better, they suggested.

By the late 1970s, the radical heat of the previous decade was cooling as a nationwide conservative movement coalesced, made up of fiscal conservatives, religious fundamentalists, regulations-averse industrialists, and libertarian-minded foes of government. Despite their differences, most shared an animosity against the recent political, social, and cultural gains achieved by anti–Vietnam War protesters, civil rights demonstrators, feminists, America's Indigenous populations, disabled individuals, and consumer activists, all of whom were suspected of having led the country in the wrong direction and now had to be countered and reversed.

At least that apparently was the view of Lewis Powell, a corporate lawyer from Virginia. In August of 1971—just a few months before he was sworn in as a US Supreme Court justice—Powell sent a confidential memorandum to the US Chamber of Commerce warning of an "Attack on the American Free Enterprise System." The troublemakers, he wrote, were, "not unexpectedly, the Communists, New Leftists and other revolutionaries who would destroy the entire system, both political and economic."[1]

However, argued Powell, "The most disquieting voices," came from "the college campus, the [non-conservative] pulpit, the media, the intellectual and literary journals, the arts and sciences, and from politicians." Thus, the time had come, according to Powell, for America's business leaders and conservatives to unite and push back against the disruptors of the 1960s. He also revealed he had a plan to achieve all this.[2]

According to a popular account, large American corporations responded to Powell's jeremiad by organizing a powerful array of lobbyists, conservative media organizations, political campaigners, and academics to oppose a leftist onslaught and lay the foundation for the modern conservative movement.

Jason Stahl, an independent historian in Minneapolis, however, thinks Powell's influence played out differently. The real impact of the memo, he argues, is that it spurred the center-right American Enterprise Institute to convince the think tanks it was already funding to counteract anti-business ideas flowing from left-leaning policy institutes with a robust pro-business narrative of their own.[3]

The conservative institutes soon obliged. And many drew "their inspiration from the successful tactics of the tobacco industry to thwart any restrictions on their activities," writes Jean-Daniel Collomb, a scholar at the Grenoble Alpes University in France. (Among his academic specialties is the ideology of climate change denial in the United States.)

These organizations also funded "seemingly independent think tanks and hired experts and scientists in order to discredit scientific research and evidence likely to justify governmental regulations on their activities." Among their main targets was government regulation of industry.[4]

Manufactured doubt appeared everywhere, taking the form of "alternative" scientific studies, books, articles, commercials, op-eds, and radio and television talk programs. Industry-backed speakers and "experts," including for-hire scientists and other academics, made themselves available to testify at hearings and complain about the environmental movement and government regulation. Much of the output flowed from the Competitive Enterprise and Heritage Foundation. According to one study, of the 141 English-language environmental books published between 1972 and 2005, more than 92 percent were linked to conservative think tanks.[5] The message was consistent and straightforward: private enterprise and freedom were under attack. Too much government regulation was unnecessary, intrusive, and expensive, killing jobs and forcing companies to move overseas. Even worse, allowing government to play such a significant role in the economy would pave for an even more oppressive form of government—socialism.

Such rhetoric resonated with many conservative Southerners, coming as it did during the ongoing Cold War between the United States and communist, authoritarian Soviet Union. For one thing, the word "socialism" was layered with extra meaning in the South. In the nineteenth century, antebellum enslavers had hurled the word as an epithet at abolitionists. In more recent times, conservative white Southerners deployed it to smear anyone advocating racial equality for Southern Blacks.

Denialism Becomes Doctrine

By the early 1980s, a national conservative movement was making big gains. This was made clear when former California governor Republican Ronald Reagan—once viewed by pundits as too far to the right to be electable—took the helm of the nation in 1981 and tried to steer it on a hard right trajectory. "The nine most terrifying words in the English language are," the new president had once quipped earlier, compressing his governing principle into a memorable, bumper sticker simplicity, "I'm from the government and I'm here to help." Conservatives cheered from coast to coast.[6]

In Reagan's eyes, current environmental regulation was an example of egregious federal overreach in need of being lassoed. In fact, "Mr. Reagan seemed to have selected the nation's environmental policies as a prime target of his social revolution," wrote *New York Times* correspondent Philip Shabecoff.[7]

This was apparent when the Reagan administration weakened existing environmental regulations and eased the way for private companies to gain access to coal, timber, and oil on federal lands. Funding was cut for federal agencies responsible for enforcing conservation and pollution laws, such as the Department of Interior and the EPA.

Reagan's efforts, however, ultimately proved less damaging than many critics feared. Although he was hardly a warrior for climate action, he received praise from environmental groups for supporting the Montreal Protocol in 1987, which gradually terminated the use of certain chemicals that depleted the ozone layer in the atmosphere, which is related to climate change.

Moreover, climate change was not yet anathema to the GOP as it is today. In fact, in June 1992, Reagan's successor—his former vice president and Texas oil man, George H. W. Bush—attended the UNFCCC meeting in Rio de Janeiro. There, Bush along with delegates from 153 other countries pledged to work together to reduce dangerous GHGs. "The United States fully intends to be the world's preeminent leader in protecting the global environment," Bush proudly announced at a news conference.[8]

Not long afterward, however, he backed away from his pledge, reportedly convinced to do so by his Chief of Staff, John Sununu, a mechanical engineer who dismissed the case for climate change as "poppycock." At the time, conservative think tanks such as the Cato Institute and the Competitive Enterprise Institute were also pressuring Bush to downplay its importance.[9]

Bush's retreat was significant. By this time, the GOP was waffling over how to respond to the growing climate concerns. Should Republicans agree to

GHG regulations? Or should they proclaim their trust in market forces and shield big donors—who were often polluters—from government scrutiny? "This issue could have gone either way, but we know which way it went," says Jerry Taylor, president of the Niskanen Center, a Washington, DC, policy institute advocating the use of market-based solutions for environmental problems.[10] As the evidence mounted daily that anthropogenic climate change was a real threat, not a hoax and not a socialist plot, the harder it became for deniers to dismiss it. And that's when "non-denier deniers" got busy.

They'd been honing a new response ever since James Hansen, a NASA scientist, testified to the US Senate on June 23, 1988, that the greenhouse effect was undeniably changing the climate. The point was to *delay* climate action by undermining the credibility of climate scientists until the climate action movement ran out of steam.

Soon climate activists found themselves accused of overstating the problem. There wasn't much anyone could do about solving climate change anyway; so, why waste time and money? Trying to do so would only embolden big government to snatch away our freedom and would dull the US competitive edge. Americans would lose their autonomy and turn it over to international committees whose members included hostile nations. Complying with internationally agreed carbon reduction rules wasn't fair either, especially if the United States were to adhere to them but the BRICS with the rising economies (Brazil, Russia, India, China) didn't.

Delayers also argued that diverting billions of dollars from the defense establishment to fight climate change posed national security implications. This, however, failed to acknowledge that every year since the 1990s to 2024, the US Department of Defense and Intelligence Community published reports identifying climate change as a national security threat. (In 2025 the second Trump administration abolished the annual reports, calling them "woke.")

Another problem for deniers was that it was hard to ignore the reality that the US national security community, the IPCC, and other prestigious scientific bodies had been correct about climate change during those same three decades. Doing so would be an admission that deregulated markets were not self-regulating and caused disastrous environmental and health problems.

So, they simply dug in and yielded nothing.

Doing so did the denial movement little harm. Millions of Americans by now merrily scoffed at climate activists as they had at an earlier generation of environmentalists. There was also another benefit to be relished. "[T]he resilience of the climate change denial movement in the face of mounting scientific evidence also highlights the weaknesses of their proponents' own ideology—a reluctance to be more straightforward about the major cultural and behavioural changes that would inevitably stem from more ecologically-sensitive climate policies," argues Jean-Daniel Collomb.[11]

In 1991 the GOP's own anti-environmental stance grew more rigid with the collapse of the Soviet Union. For years, the menace of Soviet Communism that had glued the conservative coalition together so reliably for so long was gone. GOP leaders fretted. What would bind them now?

A solution was soon on the way. "After the 1991 fall of the Soviet Union, the conservative movement replaced the 'Red Scare' with a new 'Green Scare' and became increasingly hostile toward environmental protection," concluded the authors of a 2013 report published in *Social Science Research.* Such polarization "likely will inhibit the further development and implementation of environmental policy and the diffusion of environmentally friendly behaviors."[12]

Denialism Arrives in the Oval Office

In 2000, Bush's oldest son, George W. Bush, a former Texas governor, took a turn at the American presidency. Unlike his father, though, the younger Bush did not pose as an environmentalist when he campaigned for the Oval Office. Instead, he promised "a new era" during which he intended to hand over substantial regulatory authority to states and local communities.

He kept his word. Once in office Bush moved swiftly to undermine federal authority by reducing standards for drinking water and greenhouse gas emissions from power plants. His timing, perhaps, was fortuitous because it came at a time when the environmental movement seemed to be losing public support. A 2001 Gallup Earth Day Report, in fact, revealed "most Americans are sympathetic toward the environmental movement, but are not so concerned at this time that they support drastic action to forestall potential environmental problems."[13]

Next, Bush sidelined climate action as a national priority by pulling the country out of the negotiations for the Kyoto Protocol—signed by President Bill Clinton in 1999 but never ratified by Congress—an international treaty aimed at reducing greenhouse gas emissions. He argued this measure was necessary because honoring the treaty would hurt the American economy, and besides, climate change was "not based upon science."

It was a controversial statement. In fact, shortly after taking office, Bush had asked the National Academy of Sciences (NAS) for an assessment of the IPCC's finding on climate change. Did it really exist? Were humans to blame? The academy responded unequivocally: Yes and yes. The nonprofit American Geophysical Union, the world's largest organization of earth and space scientists, agreed with the NAS.

"Yet Bush administration spokespersons continue to contend that the uncertainties in climate projections and fossil fuel emissions are too great to warrant mandatory action to slow emissions," charged the Union of Concerned Scientists.[14] Bush seemed unimpressed. "We do not know how much our climate could or will change in the future," he said after receiving the report. "We do not know how fast change will occur, or even how some of our actions could impact it."[15]

However, many other top Republican leaders still considered climate change a grave danger and supported a strong federal response. In fact, by the end of Bush's tenure as president, the GOP's 2008 party platform included a section titled "Addressing Climate Change Responsibly." A sampling of the document reads: "The same human economic activity that has brought freedom and opportunity to billions has also increased the amount of carbon in the atmosphere. While the scope and long-term consequences of this are the subject of ongoing scientific research, common sense dictates that the United States should take measured and reasonable steps today to reduce any impact on the environment."

And the best remedy, according to the platform, was "technology-

An oil refinery emitting carbon pollution directly into the atmosphere in Corpus Christi, Texas, along the Gulf Coast. iStock.com/ RoschetzkyIstockPhoto.

driven, market-based solutions" to reduce emissions.[16] There was also sharp language to placate skeptical ideologue party members: "We can—and should—address the risk of climate change based on sound science without succumbing to the no-growth radicalism that treats climate questions as dogma rather than as situations to be managed responsibly."[17]

This two-headed approach, however, weakened in the coming years, as more Republican leaders converted to denialism. Finally, many Americans were astonished and saddened to watch a handful of GOP leaders who once championed the fight against global warming themselves, such as Newt Gingrich, Mitt Romney, and John McCain, now dance to a different tune.

Orwellian thinking also arrived. Here and there it became taboo to say or write the words "climate change." In 2007, James Hansen, the NASA climate scientist, alleged the George W. Bush administration attempted to silence him from warning the public about the urgent need to act on climate change. From Florida came similar accusations that Republican Governor Rick Scott had forbidden state employees to discuss the topic in official reports.

The seas kept rising all the same.

Denialism enjoyed a big boost with the stormy arrival of the Tea Party, a movement even more conservative than the GOP itself. High on its leaders' agenda were plans to reduce both the national debt and the federal budget deficit through decreased government spending and cutting taxes. And climate change? A 2011 Yale study revealed that only 34 percent of Tea Party members surveyed "believed in" global warming. "For us, it is all about the rights of the individual over the collective," wrote former House Majority Leader, Texan Dick Armey, and conservative activist Matt Kibbe, architects of the 2010 Tea Party Manifesto.[18]

These views were not new. They reflected long established principles in existing GOP doctrine. The Tea Party's overall racial outlook, however, may have been more extreme and more Southern. According to a 2012 national survey conducted by the Diane D. Blair Center of Southern Politics and Society at the University of Arkansas, the Southern Tea Party made up 56 percent of the Tea Party national membership; however, at the time, the South—defined as the 11 states of the former Confederacy—made up only 31 percent of the American population.

"Our most striking finding in that 2012 data is that racism significantly predicts Tea Party membership in the South—and not just any kind of racism but 'old-fashioned racism,'" according to Angie Maxwell, an associate professor of political science at the University of Arkansas and director of the Blair Center. And much of this racial animus, she observed, was "anti-Obama sentiment among whites [that] has brought old-fashioned racism back into the public arena, and the Tea Party in the South gave it a home."[19]

Such sentiment was hardly confined to the South. That's because in recent decades migration patterns, cultural diffusion, and the South's growing economic clout had "Southernized" much of the country. Among the first scholars to notice this phenomenon was Arkansas-bred John S. Jackson, a political scientist and former chancellor at Southern Illinois University

Carbondale, who told a gathering at the Arkansas Political Science Association Convention in February 1986, "The South did rise again! Southern politics are now the nation's politics. We may have lost the 'War between the States' but we clearly won the long-term war for political dominance."[20]

Over the past half century, the GOP, once the party of Lincoln, too became Southernized, thanks in part to its success in attracting dissatisfied Southern democratic voters by appealing to white grievance and suspicion of federal overreach. Thus, "Southernization" had allowed Southern conservatism to not only leave its mark on the GOP within and outside the South, but also to ultimately dominate the party. "Certainly, resentment politics which Trump is the master of is closely related to race and southern politics," Jackson later observed in 2025. "The special southern Culture and Way of Life, The Lost Cause, resentment toward Reconstruction up through the Civil Rights Era, Black Lives Matter are all related and have gone national now."[21]

And that national outlook may have taken a toll on climate action.

The Tea Party's Impact

In fact, in 2012, the GOP reversed its position on climate change. Its party platform for that presidential election year had nothing to say about curbing carbon emissions, except to clarify that the Republicans opposed "any and all" cap and trade legislation to counter greenhouse gases—a market-based solution once touted by many conservatives themselves. Republicans were urged to insist that Congress "take quick action to prohibit the EPA from moving forward with new greenhouse gas regulations."[22]

Two years later, during a CNN interview, US Senator Texan Ted Cruz heartily embraced the new party line, falsely contradicting numerous scientific studies, "The last 15 years, there has been no recorded warming," he said. "Contrary to all the theories that they are expounding, there should have been warming over the last 15 years. It hasn't happened."[23]

Not to be outdone, Donald Trump, speaking to a crowd gathered at Hilton Head, South Carolina, in 2015, had this to say about global warming: "It's a hoax. I mean, it's a money-making industry, okay?"[24] His statement was a stunning reversal from a sentiment he had endorsed in a December

2009 full-page ad in the New York Times calling upon President Obama and the US Congress to take aggressive action on climate change and invest in a "Clean Energy Economy." At the time, Trump and other business leaders who signed the ad seemed worried: "If we fail to act now, it is scientifically irrefutable that there will be catastrophic and irreversible consequences for humanity and our planet."[25]

By now climate change denial was GOP creed. It was also unique. "Of all the major conservative parties in the democratic world," opined liberal pundit Jonathan Chait in 2015, "the Republican Party stands alone in its denial of the legitimacy of climate science. Indeed, the Republican Party stands alone in its conviction that no national or international response to climate change is needed."[26]

Republicans, in fact, were now saying to the world, "We aren't the *kind of people* who believe in climate change," observes Anatole Lieven, a British author, journalist, and policy analyst who has lived in the American South and written about both it and climate change.[27]

In fact, in 2016 the GOP's official position on climate change had become unmistakably hostile and bore a new Trumpian attitude: "Climate change is far from this nation's most pressing national security issue. This is the triumph of extremism over common sense, and Congress must stop it."[28] In that same document, the GOP demanded "an immediate halt to US funding for the UN's Framework Convention on Climate Change," ostensibly because US law prohibits tax money being spent on any organization that "grants Palestinian membership [in the UN] as a state."[29]

During 2020 the GOP produced no new party platform at all. There were other indications, however, of how Republicans were expected to respond to climate change. In that same year, Trump, as president, led by example by withdrawing the United States as a participant from the 2015 Paris Climate Accord, an international agreement forged by its signatories to limit global warming to 2°C above preindustrial levels.

He had many kindred spirits. The next year a study for the left-leaning Center for American Progress (CAP) Action Fund revealed that 139 out of 535 members of the 116th US Congress identified themselves as deniers. As a group, they accepted more than $61 million in lifetime contributions from the fossil fuel industries.[30] By June 2024, the number of deniers had dropped a bit to 124, reports the CAP. Meanwhile, their state legislature allies in the South have continued pushing back against climate action.

In recent years, being a climate denier, however, has put many of these lawmakers staggeringly out of step with most of their own constituents, including majorities in every southern state. Despite all the denial messaging coming their way, many Southerners couldn't help but notice all those extreme weather events in recent years. According to an October 26, 2020, study by Resources for the Future, an independent, nonprofit research institution in Washington, DC, most southern respondents said they not only believed the earth was warming, but also that governments should do more to address the problem. A 2021 Yale Climate Opinion study reported that most respondents in every southern state were convinced climate change was real and that government should treat carbon dioxide as a pollutant. However, a slim majority in Louisiana, Oklahoma, Alabama, and West Virginia agreed that humans were not responsible for that CO_2.

Denialism took a beating, or should have, when it became clear that even the fossil fuel industry had never believed what it had been saying about climate change. According to an April 2024 report, prepared by the Democratic staff of the House Committee on Oversight and Accountability (House Oversight) that drew upon investigative reporting by organizations such as *Inside Climate News* and the *Los Angeles Times*, "documents showed for the first time that fossil fuel companies internally do not dispute that they have understood since at least the 1960s that burning fossil fuels causes climate change and then worked for decades to undermine public understanding of this fact and to deny the underlying science."[31]

The staff report was specific. It zeroed in on ExxonMobil Corporation (Exxon), Chevron Corporation (Chevron), Shell USA Inc. (Shell), BP America Inc. (BP), the American Petroleum Institute (AI), and the Chamber of Commerce (the Chamber).

Despite revelations such as these, few Republican officeholders today dare oppose the party orthodoxy on climate change and incur the wrath of Donald Trump. When Trump regained the presidency in 2025 and waged war on existing climate change projects both inside and outside government, GOP reluctance became silent obedience.

Though no Democrats made the CAP 2021 denier list, their party has not escaped criticism. Two years earlier, Branko Marcetic, writing in the *Jacobin,* a socialist magazine, complained, "It's the denial of well-meaning liberals who would never be caught dead calling climate change a hoax, but scoff at the radical steps needed to meet the emergency; the denial of those who would rather plug their ears and not have to think about the problem than to really face it head on; the denial of politicians terrified that simply uttering those two words will lose them votes."[32]

Other critics fault progressives for blending climate action into a mix of other liberal goals—universal healthcare, identity politics, or defunding the police. Doing this, warns the political science analyst Anatole Lieven, has "alienated conservatives who might otherwise have recognized the threat of climate change to their nation, to democracy, and the capitalist order."[33] And many of them are troubled by climate change; they just don't know who to listen to.

In fact, "Most Republicans still do not regard climate change as a hoax," Republican strategist, Whit Ayres, told the *New York Times.* "But the entire climate change debate has now been caught up in the broader polarization of American politics."[34]

If Ayres is correct, legions of Americans who know better remain loyal to their tribe, even as their leaders not only refuse to take climate change action themselves but also work hard to undermine the efforts of others. In the American South, state governments have mastered the art.

EIGHT
YOU'D BETTER HAVE A PLAN

Texans everywhere who lived through it knew they'd talk forever about the historic frigid four-day catastrophe that crippled the operations of normal life across the state. Uri, as it was called by the Weather Channel, swept in from Canada on February 11, 2021, coating the Lone Star State with ice, snow, and deep-freeze temperatures. North Texas suffered its coldest day in 72 years, while thermometers in the Dallas-Fort Worth area registered -2°F.

The sprawling Texas power grid also failed. During the dangerous ordeal, 4.5 million Texans—almost 60 percent of the state population—lost electric power from downed lines or controlled outages when demand spiked. Worse yet, Texas utilities' backup energy was woefully inadequate. As a result, millions of people shuddered in freezing homes, communication systems fell silent, medical devices in hospitals and water pumping stations stopped working. Even the equipment at power stations froze over.

When the Texans thawed out, they learned Uri had left in its wake over 250 deaths statewide and anywhere from $80 billion to $130 billion in storm-related costs, according to the Texas State Comptroller. In February 2020, FEMA released federal relief funds to Uri's victims in all 254 Texas counties to help them recover from a winter storm that was possibly made disastrous by climate change.

Climatologists, of course, were quick to explain that while no single climate event is proof of climate change, it is possible, even if it seems counterintuitive, that warming temperatures, under the right conditions, could trigger catastrophically cold events. In Uri's case, warmer-than-usual winter temperatures, scientists hypothesized, may have allowed cold air to escape the normal restraints imposed by a phenomenon at the Arctic Circle known as the polar vortex.[1]

Blame Game

Academic discussions about climate change, however, were not a high priority in Austin, where state legislators huddled to decide how to respond to the devastating winter storm. In the end, they agreed to winterize the power grid. What they didn't do was create a *climate plan* as a safeguard against future extreme weather events. In fact, despite the severity of Uri's impact, some Texas officials instead tried to scapegoat the state's burgeoning solar and wind-power industries. Texas Governor Greg Abbott, for instance, claimed frozen wind turbines were the major cause of the freezing blackouts. "This shows how the Green New Deal [a proposed economic system designed to tackle climate change] would be a deadly deal for the United States of America," he declared. "Our wind and our solar got shut down, and they were collectively more than 10 percent of our power grid, and that thrust Texas into a situation where it was lacking power on a statewide basis. . . . It just shows that fossil fuel is necessary."[2]

Former Texas state governor Rick Perry, a former energy secretary in the Trump administration, also took a swing at the green menace and appointed himself the voice of the people. "Texans would be without electricity for longer than three days to keep the federal government out of their business," he insisted.[3]

Several state lawmakers even attacked the same solar and wind companies they had welcomed to the state a decade earlier. They did this with a raft of bills designed to hobble the alternative energy companies with onerous and expensive layers of regulation not required of fossil fuel–burning power plants. (Although these largely failed in becoming laws, lawmakers would reintroduce them again in coming years.)

State energy experts have pointed out the obvious. It was true, they said, that the state's fledgling "green" industries—which at the time produced a fraction of Texas's total energy output—did have problems operating during the storm, but so did all of Texas's power plants, including those fueled by nuclear power, natural gas, and coal, and they were the ones supplying most of the electricity. Texas's frozen natural gas lines, according to the experts, along with repeated past failure of the legislature to invest in winterizing the entire power grid and an inadequate backup power system played a much bigger role in causing the winter suffering of so many Texans.

Politicizing the facts surrounding the Uri event disturbed several top-level state energy officials. "It is . . . essential that policymakers rely on independent expert analysis rather than the moment's talking points from interested parties," reprimanded Peter Cramton, an economics professor at the University of Maryland and the University of Cologne, in his written critique of the Uri mess. As a member of the board of directors of the Electric Reliability Council of Texas, or ERCOT, the nonprofit company that ran the state's grid, Cramton was in a good position to render professional advice on this matter.[4]

Not long afterward, Cramton, along with four other board members, resigned, after being criticized by state officials for not having the grid better prepared for severe icy conditions. Nonetheless, Peter Cramton had more say: "As the world confronts climate change, our dependence on electricity must increase. . . . At least for the next 20 years, as we rely more on renewable resources for our energy, we must rely even more on natural gas resources and price-responsive demand for our reliability and resilience." During those coming years, he advised, Texas needed to modernize its electricity market design, "as the market rapidly changes technologies in response to efforts to mitigate climate change."[5]

Texas officials, meanwhile, were quick to admit their state wasn't prepared for an outlier like Uri. But that didn't mean they were ready to acknowledge that climate change played a role.

Nor were they likely to agree it had anything to do with other weather catastrophes in the state. Others, however, did notice. "Between 1980 and 2022, 44 percent of all billion-dollar climate and weather disasters (154 of 348 events) impacted Texas," reports Public Citizen, a nonprofit consumer advocacy organization. "These catastrophic events include Winter Storm Uri (2021), Hurricane Harvey (2017), and the severe drought of 2011."[6]

There was yet more bad news. In recent years, two-thirds of Texas experienced increasing amounts of flood-causing rainfall. Gulf water levels kept rising. The Texas deserts, meanwhile, were expanding. And it has been getting hotter and drier across the state.

In the fall of 2022, higher temperatures, according to water experts, were drying up Texas soil faster than normal, thus exacerbating the state's worst drought in recorded history and burdening the state with water shortages. State water managers took note of past droughts as benchmarks to help them with their planning but overlooked climate change's possible future impact on the hydrology of the state. "Climate change is unfortunately very political in Texas, and so the board cannot [plan for climate change] because of those politics," Robert Mace, the executive director and chief water police officer for the Meadows Center for Water and the Environment at Texas State University told the *Texas Tribune*.[7]

Public Citizen deemed Texas the nation's most vulnerable state to climate change (though others would give this dubious honor to Florida, or maybe Louisiana). Katherine Hayhoe, the distinguished atmospheric scientist, even thinks "Texas receives more extreme disasters that any other state"—and the United States more than other country—because both have such big portions of many *kinds* of disasters, such as floods, hurricanes, droughts, tornadoes, and so forth, all most likely amplified by climate change.[8]

Yet, for years, Texas conservative elected leaders have largely shrugged off what Hayhoe and other scientists have had to say. Such stubbornness, however, isn't unique to Texas. In fact, it puts the state legislature in league with most southern states where the idea of serious planning for climate change has yet to germinate.

AWOL in the Southland

For years, climate activists have called upon governments, big and small, to create comprehensive *climate action plans* to prepare for what's coming. This means, in brief, detailed blueprints informed by climate science and achieved with public participation that set up processes for inventorying current carbon emissions in various communities and then taking steps to reduce those gases, while also implementing adaptation and resilience measures.

By December 2022, 33 states across the country had released such various all-inclusive plans, with one state updating its document, according to the Center for Climate and Energy Solutions—or C2ES—an environmental nonprofit organization based in Arlington, Virginia.

There wasn't much to see in the South, at the time, except some window dressing here and there. Arkansas had its old 2008 final "report" to the governor with "recommendations" on how to cut emissions. Florida's "Energy & Climate Change Action Plan" offers policy recommendations "meant to highlight important energy and climate change issues, and spark discussion around these topics." Texas has come up with a "Coastal Resiliency Master Plan"—one that recognizes "The Texas coast . . . is vulnerable to an array of coastal hazards, such as coastal erosion, sea level rise, coastal storm surge, habitat loss and degradation, water quality degradation, and other issues that are putting the environmental and economic health of the coast at risk."[9] According to the master plan, the nature-based "drivers" (not human ones) contributing to this problem are "erosion, flooding and storm surge, sea level rise, subsidence, depletion of freshwater inflows and sediment deficit."[10] Oklahoma, Mississippi, Kentucky, Alabama, Arkansas, and South Carolina all lacked comprehensive plans at the state level.

North Carolina—where experts expect future hurricanes and rising seas to flood numerous toxic superfund sites and bacteria-laden hog and poultry operations that already pollute the state's air and freshwater supplies—did come up with a plan. The governor at the time, Roy Cooper, also ordered state and local agencies along with businesses to include environmental justice issues in their policy decision-making. And he called for a statewide greenhouse inventory as part of a massive effort to reduce the state's GHG emissions and to incentivize North Carolinians to put 1.25 million electric vehicles on the state's roads.[11]

By now Louisiana's then governor, democrat John Bel Edwards, had convened a special advisory group of scientists, industry leaders, and government officials tasked with producing policy recommendations, including those on environmental equity. Fifteen months later, the group produced what Edwards called "the first statewide plan in the Deep South to reduce greenhouse gas emissions to net zero from a top energy producing state." The goal of net zero was to remove an amount of CO_2 from the atmosphere equal to what a government, industry, or other organization was at the same time emitting into the atmosphere.[12]

By this time, Virginia also had a plan teeming with strategies and goals for reducing emissions, purchasing fuel efficient vehicles, making adaptations to infrastructures, addressing health and emergency service issues, and educating the public about climate change. It was also the first and only southern state to belong to the Regional Greenhouse Gas Initiative (joining Connecticut, Delaware, Maine, Maryland, Massachusetts, New Hampshire, New Jersey, New York, Pennsylvania, Rhode Island, and Vermont), the first carbon cap and trade agreement to put a lid on carbon emissions from the power sector in the Northeast and Mid-Atlantic.

However, in June 2023, guided by a new governor, Republican Glenn Youngkin, the state exited the program, only to return in November 18, 2024, when a Virginia district court ruled the governor had acted illegally.[13]

Flouting climate action isn't the only way lawmakers in a state of denial have shown their hostility to the IPCC's relentless warnings. They have also attempted to stop, if not financially hurt, others who were rising to the challenge. By 2022, at least 15 Republican-led states, most of them in the South, were crafting legislation to limit or sever relations with any companies that adopted environmental, social, and governance policies. This included climate action plans. Texas led the parade with a law requiring agencies at state and local governments to divest funds from ten financial firms and perhaps hundreds of other investment companies because they didn't support the fossil fuel industry.[14]

Obstructionism, of course, only masks the symptoms of a growing menace. "The imperative for climate solutions is not slowing up at state and local levels," says Alys Campaigne, Climate Initiative Leader for the Southern Environmental Law Center. "Lack of federal leadership stalls progress and shifts costs and risks onto citizens, cities, and states that can't ignore the realities of a rapidly warming climate. Global clean energy demand and growth, energy economics, rising insurance rates and resource constraints are going to continue to drive progress."[15]

Money talks, too. For a while, incentives in the form of federal pots of money during the Biden administration snagged the attention of officials even in climate action–averse states. In March 2023, 46 states received three million dollars each in federal EPA grants to develop plans to show how they would make an inventory of the sources of greenhouse gases and then reduce them. The final step in the process required mayors and governors to submit their state's final plans to the EPA in January of 2025. There, officials stood ready to shovel out billions of federal dollars to those states whose plans passed inspection. However, all of that was frozen when Donald Trump returned to power, wielding an axe to all federal green programs.

Maybe it wasn't a totally lost cause. That's because much of the South (or at least those states within Southern Environmental Law Center's [SELC] orbit) had been incentivized to prepare for climate harm and were "beginning to realize its full potential to make progress on climate," Campaigne points out. "For the first time, every state and large municipality in our [SELC's] region—Alabama, Tennessee, Virginia, Georgia, North Carolina, and South Carolina—has a roadmap to reduce greenhouse gas emissions and other harmful air pollution."[16]

Whether they will ever put their roadmaps to good use, especially when the Trump administration began dismantling climate action has yet to be fully grasped.

Getting Your Act Together

Why any southern state wouldn't be taking climate action, no matter what the federal government does, is a puzzlement to Daniel Gilford, a meteorologist and atmospheric scientist with Climate Central, a New Jersey-based, nonpartisan consortium of scientists and communicators. "The South should be leading the charge against climate change because it faces some of the most devastating consequences with all its low-lying places and vulnerable people," he says.[17]

If state legislators come to their senses—or are replaced by others who have about the climate challenge—what should they do? "There are only really three possible responses to climate change," Gilford suggests, "mitigation, adaptation and resilience."[18] There's also a fourth response, he adds as an afterthought. "Do nothing. There is a response to that, too. It's called suffering."[19]

The most urgent of these is *mitigation,* which means taking measures to reduce the scale and scope of the advancing climate crisis. The Union of Concerned Scientists recommends "shutting down coal and gas plants, shifting to renewable energy sources, using energy more efficiently, using less oil for transportation, ramping up use of public transportation and electric vehicles, and using land in ways that remove carbon from the atmosphere. Boldly, dramatically cutting emissions—on the order of 50 percent below 2005 levels by 2030, here in the United States—is essential to a safe climate future."[20]

The IPPC, too, has called for drastic changes in how people work, travel, recreate, farm, buy and sell, grow and harvest food, and construct things. Such actions, however, are likely to disrupt existing economic arrangements, customs, and habits of mind and cause public resistance and pushbacks.

That's why convincing people that climate change is a serious problem is vexing for even the most charismatic and effective leaders. How, for instance, do you stress the danger of a threat that is so slow in unfolding? An eighth of an inch of sea rise each year is undetectable to human eyes. Similarly, earth's temperature has also risen by 0.14°F every decade since 1880, though the rate of warming since 1982 is more than double at 0.32°F per decade.

Experts also say evolution has prepared the human mind to better focus on the moment rather than the future—doing so, ironically in this case, enhances chances of survival. Thus, it's hard to convince people to reduce fossil fuel energy consumption *today* to prevent climate problems for the unborn. Or to tell them yes, it's true—reducing GHGS will impose hardships, such as inflation, scarcities, and market disruptions, but that's the strong medicine everyone must take.

Taking these measures would also hurt the oil, gas, and coal industries.

Among the hardest hit workers in the South affected by abandoning fossil fuels would likely include many in Texas, Louisiana, and Appalachia. Small wonder many politicians in those regions aren't sounding warning bells.

And even if climate activists succeed in making their case against fossil fuels, transitioning to cleaner alternatives may also prove more difficult than some of them say it is. Not all southern locales, for instance, are blessed with steady, strong winds like those on the Texas plains. Nor is solar energy a viable alternative for everyone. The South may be soaked in sunshine, but cloud covers are also common during much of the year, making solar power production less dependable than what you would find in the arid, sunbaked Southwest, which has longer hours of sunlight every year. The Southwest also boasts a larger number of hours with direct, high-angled sunlight, the kind that maximizes energy input flowing into solar photovoltaic power plants.

Those going solar have also run into other speed bumps. Not all utilities, for instance, are keen to work with solar-minded homeowners, community solar producers, or other entities using solar technology to satisfy their own power needs and then sell surplus electricity back onto the grid. Nor do all residents welcome the arrival of large solar arrays in agricultural areas—near their favorite seacoast beaches or even on the roofs of their neighbors in suburbs—which they deem intrusive blight.

Some southern landowners, however, do welcome solar arrays because they can lease their land to utilities. Georgia Power, for instance, is paying property owners and local municipalities about $2.9 billion a year to lease their land for solar arrays. This practice angers others, however, including many Southwest Georgia residents who complain about erosion and storm runoff problems where large solar farms have been set up on sandy soils. There is also a steady fuss over the growing practice of farmers leasing a portion of their farmland to solar companies and arousing skepticism over land degradation, water rights, and loss of local food production.

Such squabbling is probably unavoidable and indicative of the problems facing communities everywhere tasked with mitigating carbon emissions. And even if they could somehow one day meet IPCC goals (which now seems improbable), huge problems remain. That's because too much released CO_2 is already embedded in the natural world. It could take centuries, if not a millennium, for it to fade.

Thus, mitigation alone isn't enough. *Adaptation* is also needed, if human life is to flourish. The term refers to adjustments people must make to respond to environmental changes already happening, and to execute them in such a way that moderates or avoids causing harm to other people and the natural world. Adaptation also means taking steps to avoid the worst blows weather extremes can deliver. Summer days are already hellishly hot and humid for many outdoor workers across the South. As temperatures rise, their employers will be under increasing pressure to restructure work schedules, provide "cool breaks," and install air conditioning in work areas in places like warehouses and delivery trucks. If these measures are not taken, workers could face heat stroke or even death. Without workplace relief, workers who are able may move to cooler places to live.

The first step usually taken in the adaptation approach is to identify emerging problems and take preventive measures, such as elevating buildings in high-risk coastal areas or even banning new developments in flood plains or wetlands and mangrove swamps. Some southern communities such as the City of Tybee Island, Georgia, are restoring mangroves and wetlands to provide natural buffers against the destruction of tropical storms. Planners must also consider which adjustments could be made incrementally to prolong or preserve the integrity of something dear to the community, such as a historic structure, an ecosystem, or a park. Other measures include encouraging people to adapt by wearing "cooling" clothes, drinking more water, and avoiding being outside during the hottest time of the day. Local government also tries to help by planting city tree canopies, building cooling centers, and incentivizing citizens to install solar arrays and evaporative or cool roofs.

Harden That Infrastructure!

Adaptation's close cousin is *resiliency*. Simply put, adaptation is the process of adjusting to new climate effects, while resilience is all about withstanding and beefing up a community's capacity to recover or "bounce back"

Dune protection efforts along the Southeast Atlantic Coast.
Photo by L. Susan Dunn.

when disruption and challenges occur. Increasingly the idea of resiliency is replacing *adaptation*, the term preferred by the IPCC. Resiliency refers to a much-needed response to climate change, and it is popular with many elected officials. For one thing, it sounds Churchillian and appeals to the human sense of resistance in the face of danger.

Resiliency, however, also allows some leaders, especially denialists, a convenient way to avoid calling the Beast by its proper name and thus escape controversy and responsibility to act. "I am not in the pews of the church of the global warming leftists," Republican Florida Governor DeSantis told reporters at one 2018 campaign stop. "I am not a global warming person. I don't want that label on me." Two years later, however, he proudly touted a one-billion-dollar "Resilient Florida" project, though he didn't mention what was causing the need for this resiliency effort. Nor did the governor focus on the fact that an infusion of federal money into the state budget helped make the resiliency project possible.

Resiliency's price tag is just one of many heart-breaking issues that at-risk communities now must confront. The residents of Tangier in Accomack County, Virginia, know this well. An island community in Chesapeake Bay, the historic fishing village with a population of fewer than 800 people has lost two-thirds of its dry land since the 1850s. Scientists expect it to be uninhabitable by the end of the century. They also say sea level rise is the main culprit, though many island residents have had trouble accepting that verdict.

Islanders do agree their community is sinking and needs help. But what should be done? Should Tangier even receive help? There isn't enough money, labor, or time to save everything, author Earl Swift suggests in his 2018 book about the island community. Moreover, there is also concern over who decides to spend the money. How much do you spend and on what? Who pays?

These same tough questions plague community leaders, planners, and citizens around the world in similar situations. That's why, say resiliency experts, all stakeholders must agree on the meaning of words and concepts before using them in policy decisions. Otherwise, opportunity is left open for those skilled at word play to leverage their interests above all others.

"Whose resilience is really being promoted?" wonders Lizzie Yarina, a designer and researcher with the Massachusetts Institute of Technology's Urban Risk Lab who studies how officials in vulnerable cities around the world approach resiliency in their urban planning. "Too often, the rhetoric of climate adaptation is for the displacement of poor, informal communities, and an alibi for unsustainable growth."

Complaints about equity arose in Miami in recent years when the US Army Corps of Engineers designed a sea wall to protect six miles of the most expensive and posh properties of downtown Miami—the sparkling sky-rises of Brickell, the city's financial district—at taxpayers' expense. Similar concerns also emerged in Charleston, South Carolina, over the Corps' plan to erect a seawall to protect the municipality's historic and commercial districts.

Lasting Sorrow in New Orleans

Stories like these are hardly new. Louisiana remains haunted by one of American's most tragic tales of the poor giving way to the wealthy. On Good Friday, April 15, 1927, after an 18-hour soaking from an ongoing fierce storm, much of New Orleans was four feet deep in floodwater. Worse yet, the city pumps that drained stormwater had been knocked out of service and still the rain fell. Here and there along its floodplain, the Mississippi River was overflowing its banks. As conditions worsened, alarmed bankers in New Orleans, worried about substantial financial investments in the city, huddled to work up a rescue plan. "Only dynamite will restore confidence," suggested one of the men.[21]

Yes! agreed the others, dynamite was just the ticket. Next, they sought help from Louisiana's political elite, who set in motion an effort to blow up a levee located 13 miles downstream to ease mounting pressure on the levees in New Orleans, even if it meant increased flooding and destruction in the St. Bernard and Plaquemines Parishes. The 10,000 people living in those communities, of course, were outraged when they heard what the moneyed elites in The Big Easy had done. "Where do they get the authority to drown us out," one of them cried, "to deprive us of our homes and living?"[22]

At last, the residents of St. Bernard and Plaquemines had to evacuate, many of them relocating to a huge warehouse. The next ten days were filled with explosions that spared New Orleans and allowed turbulent, muddy river water to flood the sacrificial parishes, causing $35 million in compensation claims from the displaced flood victims. To request, however, is not always to receive. Several corporations received reparations, but 2,809 parish residents received a pittance on average of $284 each. More than 1,000 others got not a penny.[23]

"Louisiana, Louisiana," sang composer Randy Newman in the 1970s. "They're tryin' to wash us away." His song, "Louisiana 1927" had long struck an emotional chord with Louisianans, but in 2005 it transformed into a state anthem when Hurricane Katrina flooded New Orleans, killing 986 residents, mostly poor people who couldn't escape the deadly floodwaters.

The Best Laid Plans Go Astray

Unlike the New Orleans of 1927, the city had an emergency evacuation plan in place as Katrina approached, but it fell apart even before the deadly storm struck and flooded the city, leaving more than 100,000 people trapped, including the elderly, handicapped, and poor.

Lack of foresight wasn't exactly the problem. Indeed, the city had designated its fleet of city buses to be used to transport up to 25,000 residents per day to shelters two hours to the north. But when the roads became clogged with vehicles as more than one million people frantically tried to reach higher ground, emergency officials decided to house the passengers in their care inside the huge downtown enclosed Superdome stadium instead, where other desperate and mostly poor residents were also seeking shelter. Once the storm passed, the buses could then transport everyone to the original destination. That failed to happen when New Orleans flooded so extensively that the city getaway buses were inundated. Everyone inside the stadium now had to stay put without sufficient provisions.

Critics have verbally pummeled city officials for years over the botched excavation attempt. It is not uncommon, though, for plans to go astray and the reasons are well known. Life is by its very nature unpredictable, filled with unforeseen events and circumstances and external factors we can't control. When a crisis comes, we often make mistakes and aren't flexible when things go wrong and reveal that our faith in our competencies is overblown. Only in retrospect did the New Orleans planners realize their evacuation plan was ill-conceived, poorly executed, and lacking in sufficient resources.

All these factors come into play as growing numbers of communities are realizing they'd better come up with a plan for climate change—a good one. That means their planners must rely on informed scientists telling them what to expect from climate change, a hard task because great uncertainty exists in the scientific world, not over whether we're in an anthropogenic era, but rather how bad the disruption from an increasingly chaotic climate is likely to be.

Researchers, for instance, may overlook certain vectors when they make forecasts, perhaps simply because there are so many of them. "Long term predictions based on the global signals from sea level rise are only part of the story; maybe only 50 percent," says Dr. Gary Zarillo, the Florida professor of oceanography and ocean engineering. Other forces are at work, too, he points out, such as El Niño, erosion, currents, wind patterns, terrestrial water storage, subsidence, tectonics, and the North Atlantic Oscillation—a periodic teeter-totter fluctuation in atmospheric pressure at sea level in the North Atlantic Ocean.

When Zarillo explains all this to Florida coastal planners at sea level resiliency workshops, many in the audience are surprised. For some, the biggest revelation is that in addition to anthropogenic sea level rise, there is also a "normal" rise and fall of sea level that occurs periodically. Zarillo says the level of the sea along the east coast of Florida is more than two feet lower during late summer than its seasonal high stand three months later—a significant variation. This change happens every year, getting underway during the late summer months when the northward Gulf Stream—a huge warm ocean current 40 to 60 miles wide—speeds up and causes what scientists call an "offshore directed flow" in the upper layers of the Gulf Stream. Driving this phenomenon is the Earth Coriolis Effect, a powerful spinning force caused by the earth's rotation which deflects water (and everything) clockwise in the Northern Hemisphere and the opposite direction in the Southern Hemisphere. This massive rightward push of ocean water causes a substantial drop in sea level along the Florida coast that sometimes also produces nearshore upwelling of cooler water from the lower layers of the coastal ocean. But it doesn't last long. During autumn, seasonal changes in wind patterns and cooling of sea surface temperatures in the North Atlantic, among other factors, cause the Gulf Stream to slow as much as 40 percent from its summer peak. This slowdown relaxes the Coriolis Effect, which means a rapid rebound of sea level along the coast is on the way.

How all these forces act and interact also means every coastal region, Zarillo says, has its own unique "fingerprint" that coastal planners must consider when making resiliency plans. Making calculations even tougher is that the impact of the sea level rise is amplified by melting glaciers and expanding warm waters that may one day cause the Gulf Stream to slow down—if not cease flowing—thus possibly making life even hotter in the future for residents in Florida and along the Eastern Seaboard.

And there is also the role of sand. Oceanographers have long studied the "sand budget"—the volume of sand, rocks, and other sediment washing out to sea or arriving on shore. Storm surges, the Gulf Stream, and sand replenishment projects all affect the coming and going of massive amounts of sand, a natural substance that impacts barrier islands and sea life habitats and also provides beach fronts that play an important economic role in Florida.

Not so long ago, Zarillo, who has spent 40 years doing coastal research, realized the movement of sand along beaches is "responding very rapidly" to seasonal rise and fall of sea levels that occur in a one-year period and those that are decadal. Until then, "I hadn't thought about it much before," he recalls. "I looked at my data set which was good and saw there was an inverse relationship at work." When there is more seasonal sea level rise, there is greater loss of sand; inversely, a drop in sea level rise usually indicates a gain in sand. "Again," he says, "fluctuations in the Gulf Stream play a huge role in determining variation of volume in the coastal sand reservoir, determining whether we observe beach erosion or accretion from our human perspective."[24]

Taking Stock

Good climate planning also means taking stock of those most at risk and *why* they are. It helps to have the right tools. That's why Dr. Susan Cutter and Dr. Christopher Emrich at the Hazards and Vulnerability Research Institute at the University of South Carolina developed a Social Vulnerability Index—a compilation of statistical data of a population's age, wealth, race, gender, ethnicity, and other characteristics, designed to help resiliency officers and planners better understand who is most at risk from the natural hazards amplified by climate change and where to deploy resources to better plan and respond for future disasters.[25]

The US Climate Vulnerability Index (CVI) is another wondrous high-tech

Homes along a Louisiana swamp bayou. iStock.com/Katie Dobies.

tool that can help planners and community leaders identify vulnerabilities and take steps to beef up resilience. Created by Texas A&M University and the Environmental Defense Fund, the CVI provides an interactive map of the United States enabling users to search any census tract to find out how it ranks nationally in climate change vulnerability and what is driving that vulnerability—"from quality of housing and access to supermarkets to proximity to toxic waste sites and number of deaths from air pollution," according to an EDF news release. "For example, census tracts in Houston's Settegast community rank in the 99th percentile for overall vulnerability."[26]

These indexes, along with other planning innovations, however, may have arrived too late to help some cities. Before the twenty-first century is over, scores of coastal cities in the United State, many of them southern, could be unlivable, according to a 2021 study from the Proceedings of the National Academy of Science. In fact, Miami and New Orleans may already be so impacted that remediation could be a waste of time. A more recent 2024 *Washington Post* analysis of NOAA data concludes, "One of the most rapid sea level surges on Earth is besieging the American South, forcing a reckoning for coastal communities across eight US states." The *Post* analysis reveals that from 2010 to 2023, seas along southern coasts were rising faster than in previous decades, with Gulf of Mexico waters climbing at twice the global average rate. A tide gauge at Wilmington, North Carolina, for example, showed a 7-inch rise during this period; in comparison, far to the north, Bridgeport, Connecticut experienced a 2.7-inch rise.[27]

Despite this alarming datum, too many southern legislators remain AWOL or, even worse, now brazenly trying to delegitimize climate action. In 2024, for instance, Florida lawmakers passed sweeping legislation which among other measures bans offshore windmills and makes it harder for local governments to regulate fossil fuels in their own communities. In early 2026, new GOP-initiated bills in Tallahassee appeared that would essentially neuter comprehensive climate plans for good. Similar anti-climate action measures, meanwhile, were gaining ground in at least 24 other states—half of them southern.

In the face of these reckless actions, citizens at the local and community levels are doubling down on climate issues and doing what they can to pick up the slack in responsibility displayed at higher levels of governance. Somebody must.

NINE
LOCALS LEAD WHEN THOSE IN CHARGE WON'T

Behind the august, marbled walls of the US Congress and in state capitols, elected officials often temporize instead of taking sides on any number of issues, such as climate change. The party whip usually tells them how to vote anyway. Some avoid phone calls from constituents and skip town hall meetings. The only back-home faces they are likely to see belong to campaign donors.

Life is far different, in general, for local officials. They run into their constituents every day—at church, the grocery store, or the local ball game—and these people want something done about their problems, now.

This may help explain why in the South, at least, climate action is more likely to be underway at city hall, rather than in the state capital or Washington, DC. "The human factor is often missing in the political gamesmanship in national and state governments," says Ryan Freed, Alexandria, Virginia's Climate Action manager. "But at the local level people are already suffering the effects of climate change. We know these people, and they know us. They have faces and names. They are the ones who must work in the broiling heat, or as is the case in Alexandria, face flooding events from the Potomac, which is expected to rise a few feet in coming decades. Some of our residents will have to turn off their ACs to make ends meet which could kill them. How to make peoples' lives better and safer is what I'm working to accomplish."[1]

Daniel Gilford, the Climate Central climatologist, agrees with this assessment and adds that working at the community level makes sense. "A key idea for us here at Climate Central," he says, "is to ask a lot of questions, not just explain. We want to figure out what people's values are at ground level and help local policymakers connect to those experiences and find decisions. And we make a point of talking to the 'trusted messengers' in a local community—such as elected officials, journalists, community leaders, and especially meteorologists who rank high on 'who do you trust?' surveys and already know the area. Central doesn't prescribe solutions; instead, it offers the best science and broadcast-ready graphics to help communities understand what is happening."[2]

Southern Mayors Step Forward

Mayors often rank high as those trusted messengers, and many of them were quick to join Climate Mayors, a national nonpartisan organization that made its position clear in a July 22, 2020, letter to congressional leaders: "We began the year committed to protecting human civilization and beating the ten-year clock counting down to massive, irreparable climate disruption. . . . COVID-19 has been accompanied by a profound economic crisis and a national reckoning with racial inequity faced by communities of color . . . that . . . have long suffered disproportionately from air pollution and other toxic exposure and are now getting hit hardest by the health and economic impacts of this pandemic. . . . we stand on the front lines of this moment . . . determined to meet it with strength, resilience, resolve, and bold action to protect our planet and build a more just future for all Americans."[3]

So far, the mayors then noted, Congress's work has been incomplete. Until the situation changed, mayors would lead. For starters, they pledged their cities would adopt the climate goals of the 2015 Paris Agreement to reach net zero or carbon neutrality by 2050.

Of course, success depends on what kind of city a mayor presides over. In Austin, Texas, for instance, most residents supported the Climate Mayor's agenda, says Stephen Adler, who served as the city's mayor from 2015 to 2023. "We live in a magical city that doesn't punish you for making mistakes, even when it comes to getting bank loans for the next project."

That community-wide tolerance, he adds, thrives in a city that is home to the state capitol, a university, a plethora of high-tech start-ups, and a vibrant music industry—all of which provide an atmosphere that encourages innovation and risk taking, just the values you need to carry out climate action. Austin also tends to attract progressives, who tend to support green initiatives, which in Austin, means programs such as climate preparedness and achieving net-zero carbon emissions by 2040.

In recent years, the city government sent workers across the community to solicit ideas from residents to help develop special programs fostering the development of sustainable buildings, cleaner modes of public transportation, solutions to land-use problems, and enhanced natural systems for health and recreation for residents needing such services. Adler is especially proud of Austin's Project Connect Offices high-capacity transit system expansion—a network of light rail lines, a downtown subway, an expanded bus system, and an all-electric fleet.[4]

Local Governments Going Green

Numerous other southern municipalities also have some kind of climate plan or policies. Ballotpedia, the online self-proclaimed Encyclopedia of American Politics, has tracked the nation's 50 largest cities looking for those with a climate action plan. Its 2022 listing of big southern cities that had a plan (in addition to Austin) included Atlanta, Charlotte, Dallas, Houston, Louisville, Memphis, Miami-Dade, Raleigh, Oklahoma City, and San Antonio.

These plans differ, but their overall goals generally include reducing carbon emissions, developing alternative energy systems, and taking measures to blunt damages climate change is inflicting or is likely to. For example, Miami-Dade County, which includes more than 2.6 million people on Florida's southeastern tip, consults with climate scientists, boasts its own climate staff, including resilience officials, including chief heat officer Jane Gilbert, who in 2021 became the first person in the world to claim that title. In 2017, Miami, like many other cities upset by Trump's withdrawal of the United States from the Paris Agreement, accepted the challenge to reduce its city-wide greenhouse gas emissions.

Houston, a sprawling urban mecca for oil and gas companies, has also adopted green measures. With good reason, too. "Houstonians have experienced the effects of climate change," explained Sylvester Turner, who was Houston's mayor at the time, in the opening pages of the city's climate action plan. "Hurricane Harvey was larger, slower, and had 40% more rain than it would have if it had occurred one hundred years ago. In Houston, spring arrives three weeks earlier than it did even a generation ago and our already hot summers keep getting hotter. We are adapting to climate change by raising buildings, increasing detention [temporary storage of stormwater to slow its impact on the city's drainage system], and promoting green stormwater infrastructure."[5]

Another big southern city taking climate action—and one that's tucked into the fossil fuel–rich Appalachian coal region—is Louisville, Kentucky. Among other measures, the city has taken steps to reduce CO_2 emissions by offering its citizens financial incentives to retrofit their roofs with "cool" ones—that is, roofs made of special materials designed to better reflect sunlight and heat—to reach its 100 percent clean energy goals. The aim of Louisville's *Solar Over Louisville* campaign is to empower the city government to partner with Kentucky-based Solar Energy Solutions so it can offer residents a chance to buy solar power at a discounted rate. The city has also been electrifying its municipal vehicle fleet and setting up the necessary green infrastructure to help ease the transition to battery-operated vehicles.

Other southern cities that have taken climate action, according to the green building company Elemental Green, include Birmingham, Alabama; Blacksburg, Virginia; Carrboro, North Carolina; Chattanooga, Tennessee; Clarkston, Georgia; Fairfax, Virginia; Fayetteville, Arkansas; Morgantown, West Virginia; and Newport News, Virginia.

Climate justice has also appeared within several local government agendas in recent years. "Our city started down this road in 2016 when we hired a chief equity officer whose job was to help make equity assessments for people who work in local government to help them understand the effects of racism," says Zach Baumer, climate program manager for Austin, Texas, in 2022. Since then, the city developed its plan which recognizes past discriminatory practices, such as redlining, that consigned many residents, especially Black, Indigenous, and people of color to living in polluted, industrial areas of the city.[6] Asheville, North Carolina, Atlanta, Houston, Memphis, San Antonio, and Oklahoma City are just some municipalities that have also had climate equity plans.

Strength in Numbers

Similar work is also happening at the regional level. In Georgia, for example, the Atlanta Regional Commission (ARC), an 11-county planning agency for the Atlanta region, developed the Green Communities Program in 2009 that still provides a voluntary certification process to help local governments reduce their environmental impact. So far, the ARC has certified at least 14 cities and six counties in the region. And as of 2024, the commission was also helping communities across the 29-county Atlanta Metropolitan Survey Area develop climate action plans to bolster regional resiliency.

In Florida, meanwhile, despite the state government's open hostility, the Southeast Florida Regional Climate Change Compact is on the move. Established in 2009, this powerful alliance consists of the county governments of Broward, Monroe, Palm Beach, and Miami-Dade (which include 34 incorporated municipalities, cities, towns, and villages). With its *regional* climate change plan, the climate change compact has inventoried emissions and is now figuring out how to reduce GHG emissions and create resiliency projects across South Florida, where residents produce one-third of the state's emissions.

There's no time left to debate with denialists whether this is a good idea. By the compact's own yardstick, 250,000 people in the region are living below three feet of elevation above sea level. Experts expect 21 to 54 inches of sea level rise to occur by 2070. By 2050, the compact region could be experiencing temperatures over 100°F for three months a year.

Sometimes It Takes a Village

In 2017, voters in Miami-Dade were so worried about sea level rise they did something almost un-American—that is, they voted to raise their own taxes to finance nearly $400 million worth of projects designed to lessen the impact of climate change with new measures to control flooding and fund affordable housing.[7]

Three years later, a similar vote took place in a picturesque village on Key Biscayne, an island separated from Miami by a busy causeway. There, the community of 1,700 residences, upscale hotels, and retail stores, properly called The Village of Biscayne Bay, used the ballot box to authorize issuing a $100 million bond for resiliency funds to protect local real estate and stave off dropping property values on one of the most vulnerable sea-kissed areas of Florida.

What precipitated the vote was a 2017 study commissioned by The Village itself that warned, "By 2045, if no adaptation steps are taken, most roads within The Village will experience tidal flooding during king tides, as will many low-lying residential and commercial properties."[8] However, as many observers quickly noted, the alarming report did not mention the words "climate change," perhaps to avoid riling resident denialists.

Denialism, however, was exactly what Philip Stoddard wanted to vanquish in nearby South Miami, a separate municipality, where he leveraged his position as city mayor to advocate for clean energy during his time in office from 2010 to 2020. To set an example, he drove an electric car and installed a solar array on the rooftop of his house. One of his mayoral priorities was to make solar panel installations affordable for his constituents by having the city negotiate prices with the private sector. This Sun-Mayor also helped set up solar power co-ops and pushed for a local ordinance requiring new homes to be built with solar panels. He was also willing to clash with a big utility, Florida Power & Light, when it wanted to run power transmission lines through the downtown streets of the city.[9]

What made Stoddard unusual, among mayors at least, was that he was a scientist and biology professor at Florida International University. He was also well versed in climate change. In 2015, in fact, the White House appointed him to the Governance Coordinating Committee of the National Ocean Council, where he developed national policy for addressing sea level rise. A year later, *Politico* magazine included him in the *Politico-50* "guide to the thinkers, doers and visionaries transforming American politics."[10]

Miami at the edge of sea level rise. iStock.com/Divepic.

All this, he recalls, turned him into a "novelty" for the news media, and almost every day during his stint as mayor, journalists from around the world called him for interviews. This was fine by Stoddard because it afforded him a big platform for talking about climate change, a topic which the former mayor appraises as grim. In fact, he believes, sea level rise eventually will force people of the Miami area to migrate inland, and the sooner they understand this the better off they will be.

"Miami needs to borrow as much money as the world will allow to waterproof, raise roads and restructure and then let the sea do what it will," he argues. "If we wait until things become worse our bond ratings will be in the toilet. So, you have to borrow while many lenders still think there is a chance of getting a return. All you can do is to buy time. You get the most time for your money if you take action now."[11]

That sense of urgency is also widely felt in Norfolk, Virginia, a seaport municipality situated in the state's tidewater Hampton Roads region and home to 235,000 people. The world's biggest naval shipyard is also facing ever-stronger storms and higher seas. Seawater, in fact, is rising faster there than almost anywhere else in the nation, which is one reason the US Army Corps of Engineers identified Norfolk as one of nine cities in the nation with a high risk of flooding.

It is the relentless flooding, not cataclysmic hurricanes, which provides the backbone of local support for climate action, observes Norfolk's Chief Resiliency Officer Kyle Spencer. During his many community talks, Spencer tells his audience that 13 out of the 25 top flooding events in the city have occurred since 2000, with records going back to 1936. "That has seemed to get their attention," he says. "Today, in Norfolk, nobody much is speaking denial."[12]

Besides, Norfolk residents can see for themselves the constant flooding of homes, roads, and coastlines. Small wonder that city boosters claim their city is the only one in the world with a downloadable app that sends motorists texts about hazard warnings to help them avoid driving in flooded streets.

Call in the Dutch!

In 2015, Norfolk decided it needed help. So, city leaders asked David Waggonner, a founding principal of the New Orleans–based architecture and environment firm Waggonner and Ball Architects, to bring his *Dutch Dialogues* project to Hampton Road to shed light on how his firm had used expertise from the Netherlands to help other high-risk southern coastal cities, such as New Orleans, Charleston, and Tampa prepare and respond to climate change. After all, who knew better how do such things than the Dutch, a people with eight centuries of experience dealing with floods in a country situated in a delta and living under the threat of encroaching seas? In the past four decades, they had put their knowledge to work in their own country, and in recent years have been sharing it with Americans in a series of interactive workshops.

Waggonner agreed to help and summoned the Royal Netherlands Embassy, the Hampton Roads Planning District Commission, and representatives of the cities of Norfolk, Newport News, and Hampton, along with a slew of Dutch and American planners, architects, designers, engineers, landscape architects, academics, government officials, and residents to participate in a five-day Norfolk-centered forum to capture the water wisdom of the Dutch for living at sea level.

At the onset, the Dutch water specialists encouraged the Virginians to consider the importance of their community. "They wanted us to know we are a globally important region," Spencer's predecessor, Christine Morris, told the Norfolk City Council at the time. The Dutch experts, she explained, had assessed the assets of Norfolk, a low-lying settlement in a tidal basin, situated at the mouth of the Chesapeake, a strategic location with security and economic implications for Washington, DC, nearby naval facilities, Virginia, and the nation. "When we are thinking what we have to do to thrive on water," Morris added, "it's not just in our own interest, it's in the world's."[13]

Later came the experts' advice. Avoid working in silos, they said. Gather experts from a variety of fields and work collaboratively to solve problems with flooding, water quality, sea level rise, and subsidence. Work *with* instead of against encroaching water. Your design should be comprehensive, meaning it integrates all water management tasks associated with flood control, zoning codes, drainage systems, and road construction. You can't fix flooding, but you can reduce risk of water hazards by integrating good planning into innovative landscape design. Erect engineered structures along with natural ones to mitigate the impact of sea level rise. And stop ejecting stormwater to the sea when it rains. Instead, guide it to wetlands where it can be naturally filtered. Flatten canal walls so they can serve as landscaped promenades for residents and visitors. Use stored water for recreation at strategic parklands. Reconnect low-lying areas previously severed by development. Don't forget the local economy. Set up waterfront development zones and provide ways to integrate affordable housing, economic development, and mixed used economic zones.

There were local field trips, too. Workshop participants visited nearby coastlines, wetlands, and creek beds, made observations, and later during a breakout session offered their own ideas. All this helped to shape how Norfolk and its neighboring communities planned the climate action undertaking now going on in the city.

Norfolk Gets to Work

Since those days, the community has been building seawalls, creating dune systems, and installing a variety of "gray and green infrastructures." It has also joined the Resilient Cities Network, made up of nearly 100 cities and over 40 countries, which provides access to an international cache of expertise and international connections to help planners design a new coastal community that is resistant to sea level rise and extreme weather events, limits risks to humans, and innovates an economy erected on the principles of equity and fairness. "Here in Norfolk," Spencer says, "the idea of adaptation is associated with logistical concerns like controlling flooding, but resiliency also means taking on human issues related to housing, diversity, poverty, homes and community development."[14]

Resiliency efforts also brought an economic benefit to the town. So far, says Spencer, he and his staff have reeled in resiliency investment projects worth hundreds of millions of dollars. At the time of this writing, they were optimistic that even more money would be coming once a US Army Corps of Engineers $2.6 billion Coastal Resilience plan got underway. If all goes well, the massive infrastructure project will result in nearly eight miles of floodwalls, storm-surge barriers, a mile-long levee, along with pump stations and tide gates.

Not everyone is happy, though. In 2023, residents in the majority-Black communities located along the Elizabeth River—many of which were established shortly after the Civil War—were upset when they learned the Corps' proposed massive federal floodwall would not extend into their neighborhoods. Instead, these communities were slated to receive nature-based and "property-specific" mitigation measures such as living shorelines, earth berms, reefs, and elevation of 750 homes. Although the Corps insisted that the risks Southside faced simply weren't serious enough to merit the harder infrastructure, many residents in the area with its long history of pollution believed they, once again, were being shortchanged.

The Corps' plan also troubled some environmentalists who argued the Corps was focusing too much on engineering solutions and not enough on protecting and rehabilitating natural systems that were, ironically, the very kind of projects opposed by residents along the Elizabeth River. There were also concerns from the business community that the proposed floodwalls could one day impinge on future economic development plans.

So, how does a resiliency officer handle such diverse concerns, while also having to deal with the US Navy, a variety of NGOs, local, state, and federal government officials, business owners, outdoorsmen, and bureaucracies, along with spending time in countless meetings and doing mounds of keyboarding work for compliance requirement, feasibility studies, and grant proposals, all while the seas keep heating up?

"The key thing," Spencer replies, "is to be at the highest level in government as you can. For me that's been in the city manager's office. Because of that, we're able to bring everyone to the table to figure things out. It is also about embedding yourself in the work; knowing what everyone is doing and embedding yourself in the community. You can't do it on your own."[15]

The task is also easier if the federal government is present in the decision-making. For one thing, bickering over some issues often becomes moot. "If you want to bring in billions of dollars from the federal government, you have to abide by their rules," says Spencer. "When conflict occurs, you can point out to the citizens that you can only do what you're allowed to do."[16]

Spencer likes to feel he is helping to pioneer innovative ways for urban life at the edge of the sea. "I'm going to set up the city really well for the next 50 to 70 years," he says. "And the next person after me 30 years or so from now will have to do whatever is needed at that point to get to the next stage."[17]

Alexandria Stalks Emissions

Another Virginia city, 90 miles northwest of Norfolk, historic Alexandria, Virginia, is also positioning itself for the future. Situated on the western bank of the Potomac River, just seven miles south of downtown Washington, DC, the scenic, historic, red-bricked city has been experiencing repeated and increasingly frequent bouts of flooding that scientists say are caused by old infrastructure, a hodge-podge of urban designs, and possibly climate change. In 2019, the city government was so concerned about all this, says Ryan Freed, the Climate Action manager for Alexandria and the first ever to hold this position, that it declared a climate emergency. Since then, the city has been taking actions such as beefing up its stormwater infrastructure and providing a major remodeling of its historic waterfront infrastructure that suffers periodic flooding.

Another major responsibility for Freed, a former director of the Kansas state energy program, is to help Alexandria reduce city GHG emissions 80–100 percent by 2050. Part of that effort requires reforming mass transit in the community. "A couple of years ago," he says, "we made local bus service fare free all day long. This move goes a long way in helping a population-dense community become carbon free and meet equity goals at the same time. We've also recovered and exceeded ridership levels from our pre-pandemic levels. The reality today is an outcome of fewer cars, less pollution, a stronger community, and a better way of getting around, and it's easy to do."

Alexandria has, as Freed puts it, also put its money where its mouth is. So far, the city transit company has more than a dozen fully electric buses, and the city school system has five.[18]

Making History in Mobile

In Mobile, Alabama, meanwhile, Casi Callaway is leading the campaign to make her city resilient. In 2023, she became Mobile's first chief resiliency officer and then spent the first two years figuring out what she should do in the new position. This meant she was often asking for help and advice from experienced resiliency officers around the country. Another big help was the Water Institute of the Gulf, an organization founded after the BP *Deepwater Horizon* oil spill of 2010 and filled with "scary smart" writers, scientists, and activists who have helped Houston, Jacksonville, and Charleston with their resilience insights.[19]

Before becoming a city official, Callaway spent decades as an environmental activist. Her work included a 22-year stint as the executive director of the Mobile Bay Keeper, a position that kept her busy "working against polluters and trying to protect wetlands." During those years, however, she says she avoided climate change issues, not wanting to alienate deniers and sidetrack her projects.

Keeping quiet, though, didn't mean she was ignoring the danger climate change posed to Mobile, a port city on the Gulf of Mexico. As evidence of climate change disorders grew, she knew she must speak out on the issue. In fact, Callaway even asked the mayor of Mobile to establish a chief resilience officer. "The mayor might remember it happening differently," she says, "but I kept asking until he finally relented in January of 2021 and said he would create the position, but I had to do it."[20]

An enthusiastic Callaway accepted and went to work. For starters, she assessed the city's strengths, weaknesses, and opportunities with an "eye on the economy, infrastructure, community" and how climate change would impact them. That effort also included studying justice issues such as "looking at where we did red-lining in the past and what was the impact of I-65, splitting the community."

At first, her new colleagues didn't know what to make of a former environmental activist now in their midst. "One third thought I was a narc," she says with a laugh. "One third saw me as some sort of savior and the remaining one third wasn't even sure what resiliency means."

Perhaps the one thing everybody could possibly agree on was the fact that Callaway's job even existed in Alabama showed that attitudes were changing in the Deep South.[21]

Callaway, along with legions of other climate change specialists across the South, relies on a steady supply of up-to-date verifiable climate data. Providing these data in abundance are a multitude of scientists and researchers from a wide spectrum of disciplines at universities, state agencies, and independent research institutions, often operating quietly to avoid interference from state legislators in denial. They do much more than running computer modeling and viewing data and reports. Their field work often takes them to arctic regions, agricultural areas, wetlands, oceans, mountain peaks, and faraway weather stations. Also at work on campuses are engineers, architects, designers, and social scientists crafting creative and sustainable approaches and public policies focused not only on decarbonization challenges, but also on migration, justice, and economic issues arising from climate change.

Not all the climate work is done by experts. Lay people by the droves have also unfurled their own banners and are busy making trouble for the fossil fuel status quo across the Southland.

TEN
SOUTHERN TROUBLEMAKERS

What motivates someone to become an environmental or climate activist? For many, that compulsion develops gradually as they weigh the incoming mounting evidence of human-caused degradation of the natural world. Others may be spurred into action after they've read a book like Rachel Carson's *Silent Spring*, which helped launch the environmental movement. Often people become activists when they realize it is truly *their own backyard or neighborhood forest* that's the site for a new housing development, or worse yet, oil refinery.

And then there are those individuals who are recruited suddenly to the cause because they are eyewitnesses to disaster. This was the case for civil rights activist Katherine Egland. On August 29, 2005, she steeled herself against deadly Hurricane Katrina as it smashed her hometown in Gulfport, Mississippi, drowning two beloved relatives. "I call that the moment I took my earrings off," she recalls. "[A]fter Katrina I realized 'It's here.' Climate change is no longer in our future; we're seeing it and experiencing it. That was when the trajectory of my civil rights advocacy changed to focus primarily on environmental and climate injustice."[1]

Not long afterward, Egland began hosting "Climate Change 101" classes in Black churches, where she explained how rising carbon emissions cause stronger storms and rising sea levels that overwhelm infrastructure and spread disease, and how all this ties into racism and poverty in the South. She also organized a grassroots effort to end coal-burning operations at a failing coal-burning plant five miles from her house. Her activism eventually led her to becoming a chairperson of the NAACP's Environmental and

A devastated New Orleans neighborhood, post Hurricane Katrina, August 2005. iStock.com/bjunda.

Climate Justice program. In recent years, Egland represented the organization at global climate conferences in 26 countries, including the historic Paris Climate Agreement gathering in New York in April 2015.[2]

Egland wasn't the only person stirred to action by that hurricane. Winds had barely stopped howling, in fact, when volunteers by the truckloads showed up at devasted communities all over the Gulf Coast to rescue the stranded, get victims to food and shelter, clean up the mess, and seek disaster relief funds. Some, like Egland, became apostles for climate education and action in their communities. For many, this meant trying to mitigate the human assault on wetlands and habitats to reduce the region's vulnerabilities to future storms and to expose and remedy the underlying conditions that made some human lives more vulnerable to extreme weather events. In Louisiana, such undertakings led to the creation of the philanthropic Foundation for Louisiana (FFL). With help from donors such as the Macarthur Fund and the Ford Foundation, the FFL has been busy over past decades distributing grants to individuals and groups in Louisiana that work to build stronger, more sustainable communities, especially those that focus on racial, gender, economic, criminal justice, and philanthropic issues, explains Emilia Aguinaga, FFL's Climate Justice Program Officer.

So far, according to FFL's website, the organization has provided $60 million to more than 260 nonprofit organizations in 50 of the 64 of Louisiana's parishes to fund a wide range of projects. Some focus on disaster preparedness and recovery. Others provide investment capital for small business entrepreneurs of communities often overlooked by lenders. There are also loan programs for homeowners, voter registration projects, and training sessions that help prepare local leaders to effectively interact with state officials and planners on a host of social and climate justice issues.[3]

Through its public messaging, FFL is also trying to "change the narrative" by reminding the public that the state's oil and gas industry, which dominates the political arena in Louisiana, is also polluting parishes in Cancer Alley with majority-Black populations. "We want to counter the [misleading] idea that all Louisianans are affected the same by the pollution," says Aguinaga, "and so there's no inequity problem and no need for social justice." That, she adds, is just not true.[4]

A desire for justice is what spurred Catherine Coleman Flowers into service when she moved back to her native Alabama in 2001 after having lived elsewhere for many years. An educator, civil rights activist, someone who had served in both the Air National Guard and the US Air Force, she now found herself as an economic development specialist in impoverished Lowndes County, whose population was three-fourths Black.

What struck her from the onset was that this rural community—like so many others in the South—was plagued by failing septic tanks that often backed up, causing raw sewage to flow from the ground up into bathtubs, or to swamp backyards, schools, parks, and other places whose soil was hard packed clay. The stench was bad enough, but the foul mess also contaminated drinking water and contributed to the spread of disease, such as hookworm. Conditions grew unhealthier when it rained. Today, many residents worry that climate change will only make matters worse.

Flowers was also shocked to learn that local officials had threatened to evict or even jail poor people whose inadequate or nonexistent home septic systems violated sanitation codes. What seemed unjust to her was the possibility that state health authorities would prosecute the poor and mostly Black people of Lowndes County more vigilantly than they did big corporations that polluted Alabama.

And there was also the big persistent problem—the one she'd been hired to cure: "It's impossible," she says, "to do economic development in communities where there's little or no infrastructure."[5]

Faced with all this, Flowers spent decades trying to help the residents of Lowndes establish a healthier environment. Today, Alabama's state health officials now collaborate with county and federal officials in an ongoing effort to not only clean up the septic mess but to also identify the underlying economic, political, cultural, and physical barriers that prevent economic development and residents having clean air and water.

Flowers's work has also made her a go-to authority on rural health matters. She has also authored books, provided expertise at universities, delivered expert testimony to the US Congress on the nexus between poverty and disease, and participated in former US President Joe Biden's Task Force on Climate Change. Her efforts also inspired the US Department of Justice to investigate whether racial discrimination had ever played a role in the septic nightmare in Lowndes County. (In 2025, the Trump administration shut down the inquiry.)

The Stench of Pollution

It was a sulfury smell, akin to that of rotten eggs, which led to the making of another southern troublemaker, Hilton Kelly. This stench of industrial pollution was a harm he'd lived with full time in the mostly African American West Side community of Port Arthur, Texas, where he'd grown up, played football, learned martial arts, and become an Eagle Scout.

It wasn't until he escaped and traveled beyond Texas that he realized the reeking smell wasn't universal. And yet in 2000 it was still souring the air 21 years since his departure from the industrial wasteland of his youth on the Texas Gulf Coast. Everywhere Kelly looked during a return visit that year he saw signs of a community decaying under skies turned orange by a waste incinerator, refineries, and petrochemical plants. Rates of crime, cancer, and pollution were high, and employment was low.

This, he later observed, was where Texas had assigned African Americans to live during Jim Crow. Many remained simply because it was all they had ever had.

While he'd been away, Kelly had served in the US Navy and worked as an actor and stuntman in California. Now he was back home and stood at a crossroads. Should he leave and return to a cleaner life with more opportunities elsewhere? Or stay?

Kelly, like Catherine Coleman Flowers and so many others, chose to remain in his hometown and fight. A big man renowned for his energy, grit, and leadership, he educated himself in environmental matters and spent the next 25 years leading environmental justice crusades for the residents of West Side Port Arthur. His work paid off. Among other achievements, Kelly worked out a "good neighbor" agreement with Houston-based Motiva that owned a nearby petrochemical refinery which shared a chain-link fence with a girls' playground. The corporation agreed to install new equipment to reduce emissions, provide health subsidies for residents for three years, and set up a $3.5 million fund to help jumpstart new businesses in the community.[6] Kelly also founded the Community in Power and Development Association to train local volunteers how to monitor air quality to keep polluters following the law. Today, the recipient of the 2011 Goldman Prize, he has testified before the Texas Legislature and US Senate, appeared before UNESCO, and served on the EPA's National Environmental Justice Advisory Council.[7]

An Inconvenient Question

Another Texan, Marshall L. Saunders, had spent 15 years lobbying Congress to deal with issues such as hunger and poverty when in 2006 his priorities changed after viewing Al Gore's documentary film about climate change, *An Inconvenient Truth.*

What good was helping the poverty stricken, Saunders wondered, if climate change was going to ravage their fragile homes?

Soon, he was speaking at service clubs, schools, and private organizations about the causes and dangers of climate unraveling around the world. One day, during the Q&A period following his presentation at a retirement center in California, an elderly woman asked, "What should we do?"[8]

The question took the lanky, soft-spoken Texan by surprise. "There it was," he later recalled. "I felt cornered."[9]

What the problem required, he managed to say, was "hundreds of people organizing themselves, getting educated, lobbying their members of Congress and the media in a relentless, unstoppable way, yet friendly."[10]

"Why don't you do that?" the woman asked. Her question hit home. Yes, Saunders thought. Why not me?[11]

That's when his life changed. What he was good at, he realized, was lobbying and public speaking. Why not use those skills to fight climate change? A year later, Saunders founded Citizens' Climate Lobby (CCL) a nonprofit, grassroots advocacy climate change organization that zeroes in on national climate policies.

In 2019, Saunders died, but the organization he founded lives on, boasting more than 500 chapters in countries around the world, scores of them in every southern state. Today CCL's membership is not in the hundreds as Saunders originally had hoped; instead, as of last count, it is over 200,000. Its main goal is to teach citizens how to build relationships with members of the US Congress and convince them of the importance of creating the legal and economic foundation needed for national climate policy. The first step, they say, is to stop doling out billions of dollars to the fossil fuel industry in the form of tax breaks, direct subsidies, and below-market-rates access to public lands.

Instead, the lobbyists want the federal government to establish a revenue-neutral carbon tax on all energy producers to make fossil fuels more expensive. This extra cost is expected to compel fossil fuel companies to become more efficient or switch to greener technologies. In theory, imposition of a carbon tax would also reveal to the public the true hidden "environmental" costs that accrue from burning fossil fuels that are either ignored or paid for by the taxpayer. As an extra incentive, the proposal includes a measure requiring the government to return the tax revenues collected in those transactions to all Americans in the form of monthly dividends or "carbon cashbacks" that would offset costs endured by consumers as the nation transitions to a greener economy.

So far, a carbon tax is not part of federal law, but Climate Lobby has succeeded in convincing some lawmakers in Washington to introduce the Energy Innovation and Carbon Dividend Act of 2019, a bill based on Climate Lobby's own policy plan. It awaits being reintroduced and enacted.

Acting Up

Congress's failure to pass tough climate legislation until August 2022 comes as no surprise to those climate activists who've grown impatient with status quo politics and demand more radical changes. In the South, many of these discontents belong to the Washington, DC–based youth-oriented Sunrise Movement, whose members pressure lawmakers at all levels of government to help create a "Green Economy" to fight climate change. They do more than lobby, however. In September 2019, for instance, young demonstrators occupied the lobby of the office of Nashville City Mayor, Democrat John Cooper, and demanded to know why his office was not doing something about climate change.

Cooper listened and responded. For starters, he joined the Global Covenant of Mayors for Climate & Energy. Under his leadership, city staffers produced a climate action plan that set forth goals and targets for municipal operations, pledging an 80 percent reduction in annual greenhouse gas emissions from 2014 levels by 2050. The document also called for the development of solar array projects and the implementation of "green building" codes that involve carbon cutting and energy-saving efficiencies, protecting urban tree canopies, striving to achieve zero waste at its landfills, creating more parks, and overhauling the metro transportation system with electric vehicles and mass transit. In 2024, Nashville Metro government approved an action plan containing many of these aspirations and one that acknowledges that the city has experienced a rise in devastating floods, heat extremes, and tornadic activity in recent decades.

Peaceful protests like those in Nashville may seem too tame to a United Kingdom–based group, Extinction Rebellion (XR), which according to its website prefers this course of action: "Our strategy is . . . non-violent, disruptive civil disobedience—a rebellion." This view arises from the conviction that the world has less than a decade to halt catastrophic climate damage and that a sixth mass extinction event is already happening around the planet. XR claims to have a presence in 79 countries, including chapters in the South—Georgia, Texas, North Carolina, South Carolina, Louisiana, Virginia, and Florida.

So far, southern XRers have demonstrated, blocked public streets, and occupied buildings, often coordinating their actions with those of fellow XRers around the world. They have also used theater to stage "a die-in protest and funeral procession" in the city halls of Tampa and St. Petersburg. There was once a mock wedding in Atlanta between the Georgia Public Service Commission (PSC) and Georgia Power, with the Grim Reaper presiding to protest an alleged cozy relation between the regulating agency and the utility.

Miami XR activist Nick Vazquez was 23 years old when he joined many of his generation in believing time is running out for humanity and the natural world. Extinction Rebellion provided him a course of action. Vazquez, however, chose to act alone. On Earth Day, April 22, 2021, he set out on a "climate pilgrimage" across Florida. "*I'm mad as hell,*" he messaged his fellow XRers as he hiked out into the humid, subtropical warmth. "*I'm done waiting around, so I'm walking . . . from Miami to Tallahassee . . . to sound the alarm across Florida about the climate emergency. There's no turning back. Criminal government inaction and greedy corporate self-interest have trashed the futures of my generation. We may be the last—but we won't take it lying down.*"[12]

Three weeks into his trek he reported that he had run into trouble along the way, claiming people had kicked him out of several schools and churches for trying to spread the XR gospel. Vazquez ended his journey on September 17, 2021, at the steps of the capitol building in Tallahassee, where he proclaimed climate action demands that he thought Governor Ron DeSantis should take.

The governor, unsurprisingly, did not heed the young XRer's demands. Despite this failing, Vazquez succeeded in attracting media attention to climate change.

Instead of taking it to the streets, some young activists prefer to create new climate change tools. In 2017 that's what 13-year-old Will Charouhis did after Hurricane Irma struck Miami, flooding much of the city. Suddenly, his school lessons in climate change made sense and he felt he had to act. Thus began his climate project of growing and then replanting young mangrove plants along six miles of South Florida coastline, with the aim of reestablishing natural barriers against encroaching seas and winds. He also founded *We Are Forces of Nature*, a nonprofit youth-led climate organization, whose initial goal was to plant a million mangrove plants by 2030. Since he began his work, Charouhis has addressed United Nation officials and submitted testimony to the White House Environmental Justice Council. He has also received various awards, including the Congressional Award Gold Medal.

Green Micro-Targeting

"Know your target audience market" is always sound advice if you have something to sell. That's why some activists micro-target their climate messaging to specific ethnic communities. Chispa, a League of Conservation Voters program in Florida, Texas, and other states, for instance, promotes sustainability projects and other climate actions that explicitly "reflect the values, priorities, and leadership of low-income communities, Latinx, and other BIPOC (Black, Indigenous, and people of color) communities."[13] Poder Latinx, meanwhile, another group with a presence in Arizona, California, Florida, Georgia, Texas, Washington, and North Carolina, seeks to boost political awareness and voting power among Latinos, especially those who are young. There's also a campaign to help the Latinx community better understand how climate change will impact their local communities and daily lives and what they can do about it.

Justice a Bucket at a Time

For some Southerners, climate activism means collecting scientific data themselves to bolster the cases they make against polluters. You can find many of them in the Louisiana Bucket Brigade (LBB).

Its founder is Anne Rolfes, a native of Lafayette, Louisiana, a city whose prosperity, along with high pollution levels and poverty, she believes, all come from the local oil industry. And she wants those negative conditions to change. Rolfes's activism began when, as a young woman, she spent several years as a Peace Corps volunteer in Togo, West Africa. Afterward she joined the San Francisco–based human rights organization Project Underground, which sent her back to West Africa to a refugee camp in the Republic of Benin. There she interviewed and later publicized the fate of the Ogoni people who had been driven from their homes by oil extractors in the Delta region of Nigeria, which was ravaged by oil production.

Stirred by her overseas experience, Rolfes returned to Louisiana, looking for a way to challenge Big Oil in her home state. This quest led her to a "bucket" training session near Lake Charles, Louisiana. There on the streets, she saw people carrying five-gallon buckets equipped with inexpensive, handheld air quality monitoring devices that drew air into a bag tucked inside. Next, their bags were sealed and sent to a laboratory where gas chromatography–mass spectrometry equipment checked for signs of toxic gases trapped inside. Any lab results detecting harmful toxic levels gave activists the proof they needed the next time an industry spokesperson or government regulator claimed that a stinking black cloud of factory smoke blanketing a neighborhood was safe. The bucket's innovator was Californian attorney Edward Masry, whose lawsuit against the Pacific Gas & Electric Company was showcased in the 2000 film *Erin Brockovich*.

Why not, Rolfes wondered, expand the use of this device? So, in 2000 she founded the New Orleans–based LBB, an environmental health and justice organization, which over the past two decades has partnered with numerous communities in Louisiana that suffer from industrial pollution from oil refineries and chemical plants and provided them with a raft of resources and planning assistance. "Usually, a campaign comes from or-

Led by Pastor Gregory Manning, members of the Louisiana Bucket Brigade demonstrate against attempts by the Taiwanese-based conglomerate Formosa Plastics Group to gain approval to build an additional 14 petrochemical plants on the west bank in St. James. Photo by Jack Reno Sweeney.

ganized groups, and we followed their lead," says Jack Sweeney, a former publicist for the brigade, who in 2025 was managing a New Orleans City Council campaign for Pastor Gregory Manning, whose platform included climate and environmental issues. "Our MO was asking what their needs are and making connections for them whether it be with attorneys to file lawsuits, or with members of the press. We even put on marches, protests, and on occasions do things like disrupt oil and gas conferences."[14]

Donors, big and small, help fund the LBB. The brigade also receives widespread public support, says Sweeney, "because too many people are watching the state disappear; fishing villages just aren't there anymore. They all cope with the changes in diverse ways but there's a consensus that Louisiana is a threatened place. In New Orleans, we now have a saying about those places: 'It ain't there no more.'"[15]

Louisianans, however, Sweeny is quick to add, "believe their land is still worth holding on to. It's what connects their lives to something beyond themselves. When I was in high school, I heard a lot of kids say they planned to leave the state when they got older. But many stayed here despite the state's problems. That's the tissue that connects us."[16]

Stirring the Conscience

In Houston, meanwhile, Aaron Ambroso and Tiffany Jin are trying to deal with climate uses through the humanities. In early 2021, they quit their day jobs and used their personal savings as startup money for the Houston Climate Justice Museum & Cultural Center. Located in the city's East End, the museum isn't the first of its kind (those bragging rights go to another museum in New York City), but it is unique in the South.

If climate change is truly the greatest existential threat, the 30-something duo decided, why shouldn't a museum highlight its inequitable impact on human beings? Such a place would serve as a great venue for the skills they had to offer. Ambroso was an experienced museum professional, and Jin held a degree in science with graduate level training in architecture.[17]

An art supplies warehouse provided a venue for their first climate justice exhibits. Rice University supplied the second one. Since then, the museum has been growing and spotlighting climate justice, a topic about which Ambroso cares deeply. For years, he authored articles and gave talks about the need for American museums staffs to portray the negative impacts of colonialism and racism in their exhibits. Such an idea, he says, has been around since the 1980s. "George Floyd's death wasn't the impetus for starting this museum, but it helped me to understand you can't ignore these issues anymore. Too many natural and art museums in the Western World select an object for an exhibit that is out of context and do so without mentioning its association with colonialism or capitalism."

Consider, he says, a typical exhibit that displays objects handcrafted in Africa as examples of Indigenous art without mentioning the slave trade or colonialism that may have made such an acquisition and exhibit possible in the first place. In contrast, Ambroso points out, the Climate Museum strives to "re-create or reproduce museum work to make a statement about such injustices."[18]

Now and then, Ambroso, Jin, a small staff, and a few volunteers join forces with local artists and a gardener to create exhibits that display the justice work of local activists. Among their proudest exhibits are oral histories gathered from Houston's Fifth Ward—one of the city's poorest and most blighted areas—where residents for decades were exposed to creosote used as a preservative for railroad ties at a nearby railyard and then dumped into an unlined pit for disposal, which contaminated the environment. As a result, many Fifth Ward residents fear they live in a "cancer cluster." Fifth Ward experience, Ambroso believes, is unlikely to appear in a traditional museum. "The boards of major museums here in Houston," he says, "are largely comprised of members of the oil and gas industry. The narrative on those boards, of course, is a celebratory history of the oil and gas industry that doesn't account for some of the darker sides of what those industries do and the harm they've caused others."[19]

Houston's Climate Justice Museum, however, offers different stories that feature the lives and voices of frontline communities, usually those of people of color that are missing from those larger narratives. "We're critiquing what we think are the forces heading at us in an unsustainable direction," adds Ambroso. And these insights he believes can enlighten museum goers and serve as a catalyst for change.[20]

Motivated by Faith

Storytelling as a way to spread the climate change message is what Alabama native Mallory McDuff, an environmental educator at Warren Wilson College in North Carolina, provides in her book of profiles, *Love Your Mother: 50 States, 50 Stories, and 50 Women United for Climate Justice*. "What I found in the South," she explains in the online magazine for Sojourners (an ecumenical Christian media and advocacy organization that works toward social and racial justice) "were role models on the frontlines of the climate crisis: From the hurricane zone of the Gulf Coast to the urban center of Atlanta, I discovered women who knew how to use their wit, intelligence, faith, and family to fight for their connection to home and the health of those they love." These stories, she says, changed her as a "teacher, a mother, and a person of faith with deep roots in the South."

Southerners everywhere share McDuff's conviction that faith is compatible with climate action. You find it on full display, for example, at the Myers Park Baptist Church in Charlotte, North Carolina, which publicly affirms that "what impacts One impacts the Whole. We commit ourselves and as a congregation to the intertwined responsibilities for caring for creation and seeking justice for the oppressed. Given the current and future climate crisis endangering our environment, we commit ourselves with an urgent sense of calling to protect the planet for our children's children."[21]

Meanwhile, over at the Rock Spring Congregational United Church of Christ in Arlington County, Virginia, achieving net-zero emissions with energy-efficient upgrades is a shared goal for many worshippers. It's also a priority at the First Presbyterian Church of Tallahassee, Florida, which donates money to the city government's sustainability carbon fund.

In West Virginia, faith is an important driver of the West Virginia Climate Alliance which, according to its website, is "a broad-based coalition of 20 environmental organizations, faith-based, civil rights and civic organizations, and other groups focused on climate change. . . . [Members] work together to provide science-based education on climate change to West Virginia citizens and policymakers." This alliance also seeks a reduction in GHG emissions in the state, but it also wants climate justice for communities hardest hit by fossil fuel economics, and a genuine transition for "coal miners and other fossil fuel workers likely to be impacted by the transition to a low-carbon economy."

There's even been a change of heart among many who were once faith-based skeptics. In 2007, for instance, the Southern Baptist Church adopted a resolution expressing skepticism over the scientific case for climate change. Since then, however, several top church leaders have broken rank, acknowledged the crisis, and told fellow Christians they had a duty to protect God's creation.

In Decatur, Georgia, this biblical idea of stewardship, in fact, is the guiding light for Georgia Interfaith Power & Light (GIPL), an organization

Druid Hills Presbyterian Church installed 117 panels on their roof in 2023. These panels each generate 440 watts of power and offset the church's utility bills by 26 percent, producing an estimated 73,000 kWh annually. Photo courtesy of Georgia Interfaith Power & Light (GIPL).

that sees itself as "a religious response to global climate change, resource depletion, environmental injustice, pollution." Codi Norred, GIPL's executive director, believes a close reading of the sacred text for Christians makes clear how they should respond to environmental crises. "It's pretty difficult to look at scripture and come away with the idea that we can do anything we want to the earth," he says. "Even if you think that way, it's a losing proposition because if God's creation doesn't thrive, humans can't."[22]

GIPL wants Creation to thrive. That's why, working with scientists and engineers, its Green Teams have performed hundreds of energy audits for faith communities across Georgia and awarded more than $995,000 in energy-efficiency matching grant funds to help congregations invest in energy-efficiency upgrades and install solar panels to reduce carbon footprints. They've also provided "Power-Wise and Solar Wise" assessments to show how to reduce food and water waste, helped faith-based congregations plant gardens, trees, and native plants, and assisted coastal communities as far away as Brunswick County on the Atlantic coast prepare for sea rise and climate change.

When necessary, GIPL also mobilizes its members to vote or publicly object to public projects the organization deems antithetical to sound climate action, like gas pipeline expansions across Georgia. It also sends lobbyists to Atlanta during a state legislative session, as Norred puts it, "to break apart unjust policies" and persuade lawmakers to adopt more just and green alternatives instead. "We believe one role of the faith community is to hold bad actors to account. We should be taking care of our communities, no matter what our political affiliations might be."[23]

How is their message received? "We get a pretty good reception from the General Assembly at the state capitol in Atlanta," he adds, "because in Georgia there are so many connections with faith-based groups."[24]

For centuries African Americans, especially among Southerners, have relied on faith to sustain them through bad times, from enslavement to the civil rights era and beyond. Today, many Black parishioners believe their faith requires them to face up to the dangers of climate change by taking part in a nationwide movement called Green the Church, founded by Reverend Dr. Ambrose Carroll Sr., senior pastor at Renewal Worship Center, Oakland, California.

Green the Church exists because Carroll, along with many of his parishioners, had long believed the big environmental organizations weren't listening to the environmental critiques coming from people of color. So, "In 2010, we decided we had to wake up the sleeping giant that is the Black church," he says, "on the issues of environmentalism and sustainability, really believing there's never been a successful social movement in this nation without the tone and tenor of the Black church."[25]

Today, the movement consists of more than 3,000 Black congregations across the country that regularly hold summits to educate and mobilize climate action. From their pulpits, pastors sermonize that the biblical teaching requires exercising stewardship of nature. And this means undertaking green projects such as installing solar panels, doing energy audits, and planting community gardens.[26]

Christians aren't the only ones to draw inspiration from texts they consider holy. Across the South, adherents of other faiths have been similarly motivated by their own religious doctrines and traditions, such as Georgia's Jewish Climate Action Network and South Florida Muslim Federation. In New Orleans, the Greater New Orleans Interfaith Climate Coalition includes Buddhist, Baha'i, Christian, Jewish, and other faiths. Their shared mission is to "to educate, empower, engage, and equip faith leaders and communities to meet our moral, ethical, and spiritual responsibility to establish climate justice and promote care of the Earth, and all that dwell here, through a faith and reason perspective for a healthy world."

And it was an appreciation of values like these that led Jan Booher in Boca Raton, Florida, into the field of environmental justice. As director of Unitarian Universalist Justice Florida's Climate Resilience Ministry—which is part of a worldwide, non-creedal, religious denomination seeking truth and meaning, justice, equity, and compassion—she has led many climate actions at the community level. In 2017, in fact, Booher, along with others in the Boca Raton Green Sanctuary Committee, received one of the first EPA Environmental Justice awards for an outreach project that provides online climate change information in English, Spanish, and French to Haitian refugees living in the South Florida area. This is where Booher has spent much of her time, she says, "working with people of color impacted by climate change" who are usually poor and "invisible" to the rest of society. "These are people who live in the same census tract as rich Floridians on the coast, but who are terribly impacted, and often displaced by hurricanes." Booher also helped set up a system that monitors both indoor and outdoor air quality for several Central Florida African American communities living around Interstate 4, a highway that traverses Orlando. "These people are living," she says, "in our toilets."[27]

Those Who Simply Just Pitch In

Lots of climate action in the South, of course, flows from secular organizations, ranging from short-lived grassroots groups to the big ones, such as the Sierra Club, Earth First!, and Earth Justice that are powerful enough to litigate powerful polluters and lobby lawmakers.

There are also groups that avoid the limelight and quietly buy and conserve environmentally endangered properties, such as the Nature Conservancy, the Land Trust of North Alabama, the North Carolina Coastal Land Trust, and Florida's Alachua Conservation Trust.

Although overall volunteer rates in some southern states are lower than the national average, hundreds of thousands of Southerners for many reasons keep showing up to lend a hand. Across the region, they are restoring coral reefs, mangroves, oyster beds and native habitats, removing exotic plants, protecting coastal areas, swamps, rivers and springs, monitoring wildlife and water quality, planting trees and community gardens to mitigate urban heat, helping authorities with disaster response preparedness hurricane recovery, and doing what they can to advance climate justice.

But is all this admirable work enough to counteract the endless spewing of GHGS from the present fossil fuel–dependent economic system? Or to stand up to the bullying of fossil fuel interests and their political allies at all levels of government? As the famous architect Buckminster Fuller pointed out: "You never change things by fighting the existing reality. To change something, build a new model that makes the existing model obsolete."

For many climate activists that model is the Green New Deal.

ELEVEN
SOUTHERN-STYLE GREEN NEW DEALS

A good first step in combating climate change is to overhaul the electric grid, suggested Thomas Friedman, the Pulitzer Prize–winning *New York Times* columnist in a January 2007 opinion piece. That's why he called for a "Green New Deal" similar in scale to the federal government's massive response to the Great Depression in the 1930s.

"[We] will only green the world when we change the very nature of the electricity grid—moving it away from dirty coal or oil to clean coal and renewables," Friedman wrote. "And that is a huge industrial project.... [L]ike the New Deal [the federal government's response to the Great Depression], if we undertake the green version, it has the potential to create a whole new clean power industry to spur our economy into the 21st century."[1]

But how would it work? Friedman didn't explain, but others would. In 2019, during the 116th United States Congress, for instance, Congresswoman Alexandria Ocasio-Cortez and Senator Ed Markey, both Democrats, introduced a resolution requiring the federal government to create a Green New Deal, including language acknowledging the huge impact climate change was expected to make on the economy, the environment, and national security. They also presented several resolutions to Congress tied to a 10-year action plan that included ways for the federal government to implement environmental and social justice reforms, such as reducing carbon emissions to zero by 2030 and supplying 100 percent of America's power demand with clean, renewable energy. Such an undertaking was expected to create millions of jobs and help provide all Americans access to a healthy natural world, clean water and air, and a sustainable environment.

Many climate activists rejoiced. At last, the federal government seemed serious about climate action. Their hopes dimmed later that year, however, when the Senate derailed the resolution with a vote divided along party lines.

The Green New Deal idea did not vanish. Various organizations in North America, Europe, and elsewhere came up with their versions of how the idea could work. Despite their differences, most proposals shared basic core concepts that the Sierra Club sums up like this: "The Green New Deal calls for a big, bold transformation of the economy to tackle the twin crises of inequality and climate change. It would mobilize vast public resources to help us transition from an economy built on exploitation and fossil fuels to one driven by dignified work and clean energy."[2]

Unlike the European Commission, which in 2019 approved the European Green Deal—a set of policy initiatives aimed at making the European Union carbon neutral by 2050—Ocasio-Cortez and Markey's plan never became law, though many of its key ideas appeared in the Inflation Reduction Act of 2022 (IRA) passed by Congress and signed by President Joe Biden.

The IRA provided far less than what many progressives had demanded, but it was nonetheless the most significant piece of legislation tackling climate change and environmental justice in American history. It authorized spending nearly $370 billion to boost renewable energy production and reduce GHG emissions. The act also included tax credits to boost demand for electric vehicles and rooftop solar panels. There were provisions for setting up a green bank to fund, for example, the production of clean heavy-duty vehicles, and seaports with zero emissions.

Federal money was a set aside for environmental justice efforts that augmented previous Biden executive orders, along with block grants to aid disadvantaged communities responding to harmful environmental and health conditions related to pollution and climate change. Resiliency was also part of the federal act, which requires federal officials to focus special attention on tribal and native Hawaiian communities by expanding the number of green spaces and funding affordable, climate-resilient housing.[3]

The Biden administration also authorized $20 billion in grants to establish a National Clean Financing Network to incentivize a variety of public-private clean technology projects across the nation. In September 2023, Biden, using executive action, established a federal climate corps—a public works program like the Civilian Conservation Corps created in the Great Depression. Administration officials expected the new corps to employ 20,000 people tasked with conservation, restoration, resilience, clean energy, and environmental justice projects. New rules set limits on carbon pollution emissions at power plants.

Complaints Right and Left

The IRA, however, was controversial. Denialists and other conservatives hated it. All 50 Republican senators, in fact, voted against the bill becoming law, even though some of them had helped write some of its provisions. The new legislation was costly and inflationary, they argued, too intrusive into the free market, extremist, and ineffective. Some critics insisted the IRA was a radical undertaking. "It wasn't long ago that the Green New Deal was seen as so extreme that it was laughable. Well, guess what, here we are, and it isn't funny," wrote Daren Bakst, an analyst for the Heritage Foundation, a conservative think tank.[4]

Many progressives were also unenthusiastic. For them the IRA was weak. It made too many concessions to fossil fuel interests, they argued, citing as an example a provision requiring the Interior Department to lease at least 60 million acres for new oil and gas extraction on public land before it could lease public land for offshore wind development.

At least one independent group agreed with this analysis. It was Oil Change International (OCI), a global research, communication, and advocacy organization that kept track of the true costs of fossil fuels. On November 2023, OCI released a study that concluded that the IRA gave the fossil fuel industries a pass and thus failed to accomplish very much. "Overall, US domestic petroleum demand is expected to decline by 10%," reported the international team. "Despite this drop, production is expected to increase by 13% and exports by 23%. . . . As rising exports of oil and oil products fuel the climate crisis, the United States will be exporting emissions and claiming progress at home."

There was more disappointing news. It arrived in a 2024 report from the Rhodium Group which found the United States was not on track to meet its commitment to reduce emissions to be on a path to limit global warming to 1.5°C. Despite all these criticisms, the IRA so far has been the *only* major federal effort to combat climate change.

Green Economy Advocates in the South

During the Biden years, plenty of Southerners at the grassroots level were trying to implement *regional* Green New Deal measures of their own. Many were focused on the Gulf of Mexico region where, according to the US Energy Information Association, "Over 47% of total US petroleum refining capacity is located . . . as well as 51% of total US natural gas processing plant capacity." The Gulf South for a Green New Deal, a coalition of more than 350 organizations, for example, has been striving to achieve climate, racial, and economic justice in Texas, Louisiana, Mississippi, Alabama, Florida, and the Territory of Puerto Rico.

The New Orleans–based coalition formed in May 2019 and produced a policy platform. It brimmed with aspirations, policy suggestions, principles, goals, and strategies on how to make Green New Deal ideals compatible with the realities of the Gulf Area, which "disproportionately suffers many of the worst effects of climate change in the US—from increased temperatures to hurricanes and flooding, all while losing land to the rising seas."[5]

Today, Gulf South for a Green New Deal has morphed into a new organization—Gulf South to Appalachia (GS2A) representing 17 states. According to its website, GS2A is "becoming one of the largest climate justice formations in the nation."

In the spring of 2021, another regional Green Deal movement emerged when more than 160 southern organizations collectively endorsed the Southern Communities for a Green New Deal (SC4GND) policy platform. Altogether, over 100 leaders and activists from every southern state except Oklahoma gathered to build a network of Green New Deal projects and assist climate-threatened at-risk communities.

They also wanted to give their work a Southern flourish.

"The Green New Deal is a great idea for liberal progressive cities up north, but it wasn't going to work the same way in the South," says Scot Quaranda, communications director for the Asheville, North Carolina–based Dogwood Alliance, a founding member of SC4GND, "so, our people decided to stress our values and that meant stressing the importance of environmental justice and what the greater view looks like."[6]

This meant they had to consider the realities of life in the South experienced by everybody. As a result, their action plans acknowledged both the geography of the South and the region's troubled past, including the legacy of hostility from southern state governments to federal laws intended to protect the lives and liberties of minorities.

"Everyone got to throw in their two cents worth," recalls Quaranda. "We developed a platform that people in the South can use wherever they live to make a happier, greener, and more just world. We want to make sure they have a way to communicate and the resources they need that others can provide if necessary. We also want to be able to empower people to go to their elected representatives and say this what the people that you represent want you to do."[7]

This doesn't mean their members one day won't find themselves at cross purposes. Quaranda knows from experience that this kind of problem can occur, even when people claim to be working toward a common goal. For example, "The Dogwood Alliance started around 1996 to push back against the industrial deforestation of southern trees, especially along the North Carolina and South Carolina coasts, which were being mowed down and turned into wood chips for the emerging bio-energy trend," he says.

In 2010, the situation was still bad, and it was largely driven by many European nations transitioning to renewable energy and trying to cut their carbon emissions by getting carbon credits for using renewable forest products from the American South. Significant European subsidies helped pay for it all. One of the several large corporations that benefited from the tree cutting was DRAX, a big UK company that produces 5 percent of all power in the United Kingdom. The removal of those trees, which the Dogwood Alliance opposed, harmed the southern coastal environment and made it more vulnerable to extreme weather events. "It's crazy economics," says Quaranda wearily. And it was a crazy way to combat climate change.[8]

Struggling in Birmingham

In August 2021, Michael Hansen—who at the time was the executive director of Greater-Birmingham Alliance to Stop Pollution (GASP)—was leading the Southern Communities for a Green New Deal, a grassroots drawn-out crusade to convince the politicians and city officials of Birmingham, Alabama, to develop a homegrown version of the Green New Deal.

City Hall wasn't impressed. "Every time we've talked with them," Hansen recalls, "we were met over and over with the comment 'where's your plan?' Yes, they're paid to say this, but after years of hearing it, we decided, okay, you won't do it, so we will. We're trying to call your bluff; we'll give you a plan."[9]

So, GASP's staff and volunteers went to work, cobbling a "people-driven, justice-first" plan imbued with Green New Deal principles they hoped would be adopted locally and replicated by other municipalities. For inspiration and guidance, they drew upon the expertise of focus groups, leadership teams, and former government employees with policy experience. Overlooked "front line" residents were consulted, too. "In some cases," says Hansen, "we offered stipends like childcare so certain people would be able to attend the meetings."[10]

Despite its well-intentioned labor, GASP hasn't yet achieved its dream. By the summer of 2025, the City of Birmingham had adopted some climate actions, but no Green New Deal. Nonetheless, GASP, through its "awareness campaigns, storytelling, and civic engagement activities," is still in the fight.

Greening the Public Sector

The enablers of the fossil fuel industry may succeed in manipulating governments big and small in the United States to ignore, forestall, or hobble climate action. But they may have far less luck in thwarting the private sector's march to a cleaner, greener, economy.

"Let's be clear that the Great Disruption is occurring in large part, because the marketplace is speaking. Every government will have to follow the market or face the consequences," writes entrepreneur and economist Jeremy Rifkin, a senior lecturer at the Wharton School's executive education program at the University of Pennsylvania and author of 19 best-selling books, including *The Green New Deal.* Around the world, according to Rifkin—who is also a consultant to the 27-member nations of the European Union—people are decoupling from carbon and transitioning away from a fossil fuel–based economy based on technologies and business practices from the Industrial Revolution and moving to a new and emerging one that has grown from the convergence of a digital-based internet communication technology and an array of new forms of renewable energy, such as wind and solar, that people can distribute widely and efficiently and administer cooperatively.

If Rifkin is right, denialism in the United States may be hurting the nation's economic future, as other nations pivot to green energy. China, for instance, is creating a new generation of alternative energy infrastructure faster and bigger than found anywhere else in the world. It also produces two-thirds of the world's electric vehicles. In 2025, global investment in clean energy was expected to surpass that spent on gas and oil. It has become clear that demand is shifting away from fossil fuels. For the first time ever, renewable energy has surpassed coal as the primary source of electricity around the world, according to a report from Ember, a global energy think tank. Meanwhile, production costs of clean energy systems have been dropping, as the technology behind it all has kept improving.

Whether the fossil fuel industry CEOs and their Southern denialist/ allies like it or not, these changes are coming to the South, too, if not in waves, then in a rash of "green" dots of eco-economic activity. In Ector County, Texas, for instance, Occidental's subsidiary 1PointFive wants to make money by extracting CO_2 out of the air. That's the big mission of its first large-scale—and when completed the world's largest of its kind—Direct Air Capture plant now under construction. Once the project is up and running, officials expect it to be capable of vacuuming from the sky up to 500,000 metric tons of carbon dioxide per year.[11] Construction of 100

Wind farm in Texas. iStock.com/mjo007.

more such plants could be on the way in the future. The company business plan is to sell carbon credits to other companies in what could be a $150 billion market by 2050, according to Bloomberg analysts. At the moment, 1PointFive has five other CO_2 capture and sequestration projects in Texas and Louisiana.

Texas, meanwhile, also boasts more than 11,000 wind turbines in the High Plains region, which is more than any other state. It also ranks number one in producing renewable energy; in 2022, solar, wind, and geothermal powers accounted for about one-third of the state's electricity needs. By the next year, Oklahoma, producing 42 percent of its electricity from wind, was number three.

By 2025, many utilities across the South, including some that had once resisted alternative energy projects, were now adding such enterprises to their own suite of power production systems. Florida Power & Light Company (FPL), for instance, had installed 78 solar energy centers in 31 counties across the Sunshine State—producing electricity for 1.1 million homes. The company was also applying the final additions to the state's first-ever hydrogen hub—whose electrolyzers compress hydrogen blended with natural gas to fuel a combustion turbine that generates electricity. By 2031, FPL expects 35 percent of its power to be solar generated.

A revolutionary kind of power may be arriving in the South, too. In 2025, near Richmond, Virginia, in Chesterfield County, Commonwealth Fusion Systems (CFS), an MIT-incubated company, is in the planning stages for what could become the world's first commercial grid-sized fusion power plant capable of producing 400 megawatts of power and plugging into the state's electrical grid by the early 2030s.

Best of all, according to CFS, once built, the new plant will emit zero GHGS. The company also hopes it can one day help decarbonize heavy industries, such as steel, cement, and chemical, by replacing the fossil fuels they now depend on with an abundant, clean, consistent supply of electricity and extremely high-temperature heat supplied by fusion power.

All this depends on the company's ability to make power as the sun does, by creating more energy output than input on a commercial scale. Skeptics, however, such as John Holdren, codirector of the Harvard Kennedy School Belfer Center's Science, Technology, and Public Policy Program, have doubts about the feasibility of producing fusion power any time soon, citing concerns over construction and operational costs and the risk of radioactive contamination. Holdren also worries that being overly enthusiastic about fusion's possibilities "feeds the public's belief in technological miracles that will save us from the difficult task of dealing with climate change."[12]

Bob Mumgaard, the company's CEO, however, remains confident CFS is on the right track. In fact, the Virginia plant, he says, "is the first of thousands of fusion power plants we plan to put on the grid as part of our mission to deliver fusion energy with the urgency the energy transition demands."[13] That sense of urgency is also driving fusion energy companies in the United States, France, Germany, China, Canada, and the United Kingdom to compete for a share of a potentially huge global market.

However, the growing demand for such an extraordinary source of power is also a troubling sign. That's because it reflects a voracious appetite for electricity to power the numerous, huge, energy-intensive and ever-increasing number of data centers needed for the internet, cloud computing, and artificial intelligence that keep popping up in Northern Virginia (the data center of the world), Texas, Georgia, and other states. All rely on a grid powered mostly by fossil fuels. This means emissions are increasing and escaping uncaptured all the time. The arrival of fusion as a clean power source capable of operating at a scale large enough to substantially offset these emissions is most likely decades away. Given the IPCC's timetable for climate action, none of this is good news for climate change mitigation efforts.

Do-It-Yourself Solar Power

Across the South, meanwhile, other entrepreneurs are producing alternative energy that can be used now. For Tennessee Governor Philip Bredesen Jr. and two colleagues, this means making solar power. In 2011, they founded Clearloop, a business whose purpose it is to steer southern companies off the dirty energy power grid and onto something cleaner. To do this, the company uses its clients' own money to build solar farms that supply energy for those same clients.

Bredesen and his partners decided early on there was no point in setting up their operations in, say, California, where you'd only be competing with plenty of solar companies. Nobody, however, could say that the South has too many solar farms with electricity for sale. So why not make that region your primary market? And why not build the solar farms yourself?

A bigger question, though, was this: How could they accurately estimate how much carbon a client avoids by going solar? That's why Clearloop teamed up with WattTime, a California-based nonprofit that specializes in providing quantified carbon use data.

By 2024, Clearloop belonged to Silicon Road, also based in Nashville, a utility-scaled supplier of alternative power that provided experience, technical help, and economy of scale that helped the company—now a subsidiary—to grow and flourish. Today, Clearloop has solar array farms in Tennessee, Mississippi, and Louisiana, creating enough sun-power electricity to power 4,000 homes in the Southeast.

One of its clients is Nashville's Vanderbilt University, which uses the company's services to offset its own carbon footprint at the campus level. Mindful of social justice, Vanderbilt aims to help provide low-coast solar energy to 600 homes in Brownsville, a mostly Black and African American rural community in central Haywood County located not far from Memphis, a city with a history of lynchings and violent opposition to Black participation in civic life.

Building EVs

By 2023, the South was also witnessing the growth in the manufacturing of electric-powered technology. The trend was so hot at the time that many New South boosters proclaimed a "New Detroit" was in the making. Hyperbole? Maybe. But the South did seem on its way to becoming an important North American hub for new EV assembly and battery production plant projects.

By June 2025, six southern states—Alabama, Florida, Georgia, North Carolina, South Carolina, and Tennessee—were significantly involved in the EV market in some way, according to a study by Atlas Public Policy in partnership with the Southern Alliance for Clean Energy. For instance, in Tennessee, along a half-mile stretch between Spring Hill and Columbia in a facility of 2.8 million square feet, or about the size of 45 football fields, Ultium Cells LLC, a joint venture between General Motors and LG Energy Solutions, is mass-producing lithium-ion battery cells for electric vehicles. And by late 2027, the company expects to be producing lower-cost LFP battery cells as well.

These developments are part of a larger trend of industrial output shifting away from the Midwest, where General Motors and Ford are headquartered, and toward states like Alabama, Georgia, and the Carolinas, places where it is warmer, wages are cheaper, and unions weaker. Tax policies and financial incentives also play a role.

In October 2024, there was big EV news in Florida when Aura Aero,

A 19-seater hybrid-electric regional aircraft manufactured by Aura Aero in Daytona. Photo courtesy of Aura Aero.

a French aircraft design and manufacturing company, announced it was building a 500,000-square-foot manufacturing and assembly plant in Daytona Beach International Airport. Here the company will fabricate a new generation of regional hybrid-electric, lithium battery–powered aircraft, which according to its website will "enable in-flight emissions to be reduced by 80% compared to business aircraft in the same category." Company officials hope to begin construction in 2026 and ultimately create 1,000 new jobs.[14]

In the coming years, there could be ample green fuel to power those electric planes, thanks to the discovery of new southern assets. During the same month as Aura Aero's announcement, the US Geological Survey released a study that indicated between 5 and 19 million tons of lithium reserves lie beneath the ground in southwestern Arkansas. "Lithium is a critical mineral for the energy transition, and the potential for increased US production to replace imports has implications for employment, manufacturing and supply-chain resilience," says David Applegate, USGS director.[15]

Other economic factors are also in play that may lead the South to a greener future. Location, labor, material, know-how, technology, investment money, along with a growing demand for EV technology all seem to be coming together to ensure that the South will be one of many global sites now prepping for a new age of EV production.

The market trend toward alternative energy is even influencing the decision-making of large-scale land developers. Consider the ever-expanding Babcock Ranch, formerly a sprawling cattle and timber ranch, located on flat scrublands in Southwest Florida that is fast becoming a small city. In a complicated 2006 real estate transaction, developers bought the 91,000-acre ranch property that required them to sell 73,000 acres to the state government, which now manages them as a natural preserve, consisting of wetlands, greenways, and wildlife corridors.

The remaining 8,000 acres are now being developed for homes, commercial buildings, schools, medical centers, and a community center by Sydney Kitson, a former professional football player and prime mover behind the Babcock Enterprise. And yes, sea level rise, climate change, and ever-more-powerful hurricanes really did weigh on his mind during the planning stage for the community decades ago, says Lisa Hall, Babcock Ranch spokesperson and former television journalist, as she drives her visitors on a tour of the burgeoning community in her red electric Tesla. "That's why unlike other developers we didn't build along the Gulf coast. Here we're 30 feet above sea level."[16]

Babcock's engineers didn't just look at altitude before they began building. They also studied the natural layout of the land to understand features such as the natural flow-way of water, hydrology, and soil types. This knowledge allowed them to combine nature-based drainage with human engineering to produce a complex of remotely controlled "smart ponds" that track and direct stormwater through swales and green spaces to control flooding, an all-too-common danger in this scrubby, humid stretch of hurricane country.

Sustainability and resiliency plans were in the blueprints, too. That's why today all the Babcock homes (5,000 or so built so far) meet state criteria to qualify as green homes with Energy Star certification. Residential, retail, and commercial properties are all powered by a massive array of more than 30,000 solar panels (owned and operated by FPL) and potent storage battery systems on 850 acres that generate enough energy for 30,000 homes.

One of the many "solar catchers," or solar trees (panels) that allow residents of Babcock Ranch, Florida, to charge their phones and other handheld devices. Photo by L. Susan Dunn.

Right now, technicians send so much surplus solar power back into the grid that Babcock Ranch officials can proclaim their community is the first 100 percent solar-powered town in America.

"We don't do 'spin' here," Hall insists. For instance, "We really planned to be a shelter in place community when hurricanes come."[17] Nature tested this goal in late September of 2022 when Hurricane Ian blasted nearby Gulf coast communities and pounded Babcock Ranch with 150 mph wind gusts and heavy rains. In its wake, the storm left 149 deaths and $112 billion in property damage. It was the most expensive storm in Florida history. A Category 4 hurricane with 140 mph winds and 10 to 15-foot storm surges, Ian flattened sections in the nearby cities of Naples, Fort Myers, and Cape Coral.

Babcock Ranch, except for a few missing shingles and uprooted trees, however, was almost unscathed. Residents never lost electricity during the storm; nor did they endure a loss of telephone or internet service. For one thing, no aboveground power cables, telephone lines, or utility poles exist in the community.

Word of Babcock's success went viral. Journalists along with engineers and government officials from around the world rushed in to find out what made the ranch community so resilient. "Developers came, too," says Hall, "and told Mr. Kitson they wanted his playbook. He'd tell them, sure they could have it, but they've to agree to make things even better."[18]

Such a bargain should, perhaps, be mandatory for all climate-related transactions. After all, making the world better by protecting it and helping justice thrive is the goal of most climate activists, if not most good people.

In fact, there are some activists who want new rules for the climate age. That's why they keep going to court.

TWELVE
TAKING IT TO THE JUDGE

In 1987 Reverend Benjamin Chavis was delivering a speech to the National Press Club about the United Church of Christ landfill study he coauthored when his mind drew a blank. The words he needed to describe what he was up against in those days just weren't there.

Then, he later recalled: "It came to me—environmental racism. That's when I coined the term. To me, that's what it is."[1]

Chavis's words were "a transforming moment for environmental law in the United States," wrote Georgetown law professor Richard J. Lazarus in 2000. The legal codes at the time contained little on the idea of racism in environmental law and not much existed in federal regulatory rules. All of that changed when Chavis gave birth to an idea by naming it. "The environmental justice movement, propelled by environmental racism claims," Lazarus added, "has led to a significant renegotiation of the terms of environmental law in a variety of subtle but significant contexts."[2]

When Chavis wrote his landfill study, environmental law itself was still a newcomer to jurisprudence, having emerged in the mid-twentieth century. As a body of law, it now spans a wide web of treaties, statutes, regulations, conventions, and policies designed to protect the environment and human health and natural resources.

Much of what environmental attorneys do takes place in civil court, where they try to hold polluters and others responsible for the damage they cause to people, property, and the natural world. Sometimes they sue government agencies for failing to enforce environmental laws. Or, if attorneys represent a corporate client, they might challenge actions taken by an agency or the constitutionality of a law. Special Administrative Law courtrooms are where clashing lawyers argue the merits of rule-making decisions promulgated by government agencies, such as the Environmental Protection Agency, or a state department of environmental protection.

Southern Attorneys on the Move

There is always plenty of work like this to do in the South. "There are so many contaminated sites [thousands according to the EPA] across the South that will never be properly addressed without the assistance of great communities and great lawyers working together," says Chandra Taylor, senior attorney and leader of the Environmental Justice Initiative in the Chapel Hill, North Carolina, office of the nonprofit SELC.[3]

Taylor's employer is the largest nonprofit, nonpartisan environmental legal advocacy organization in the South. Its primary goals, she explains, are protecting the right to clean air and clean water, preserving the South's rich natural diversity and beloved places, and fighting against environmental degradation that too often impacts communities lacking the financial ability or political influence to defend themselves.

"We're more in the regulatory side of the issue," Taylor explains. "We're not raising tort theories, and we try to prevent environmental dangerous harms. We find out if there's been a violation of a regulation, if so then we file a lawsuit. We bring cases on behalf of other organizations whose missions are to monitor pollution, such as the various Riverkeepers organizations. We don't usually do individual cases; we represent the public."

The SELC focuses on the Southeast, with a special interest in environmental justice cases and those whose favorable outcomes could slow the progress of global climate change.[4] Taylor loves the work. She's a North Carolina native who combines a love of science, a desire to fight for social justice, and a willingness to jump into civil cases. Since her arrival at the SELC in 2016, she has learned much, especially the importance of being prepared. "We want to be armed with facts. There is something very powerful in being able to add to your argument: The law says!"[5]

Of the many environmental justice cases she's handled over the years is one she likes to showcase in public presentations that revolves around the small community of Navassa, located along the Brunswick and Cape Fear Rivers not far from Wilmington, North Carolina. From 1936 to 1974, the mostly African American community was the site of a creosote production operation owned by the Kerr-McGee Corporation on a 251-acre wood treatment facility. During those years, excess creosote settled into nearby wetlands that flow into Sturgeon Creek, and the Brunswick River. It also seeped into the groundwater, threatening humans and local wildlife.

When Kerr-McGee closed the plant, it tried to hide its assets and avoid liability for the pollution, Taylor says. "And they were not successful." The state turned the matter over to the EPA, which in 2004 was part of a $5.15 billion settlement against Kerr-McGee and other related subsidiaries of the Anadarko Petroleum Corporation. It was the biggest governmental win for environmental cleanup in US history. The money was directed into federal superfund coffers to help pay for cleanup of America's most toxic sites. Navassa is one of them.

However, there wasn't then, or now, enough federal money to decontaminate all 1,340 superfund sites (313 in the South) across the country. And even if there were, federal regulations don't require restoration efforts to reach a pristine natural condition, but rather one that allows the site to safely continue at pollution levels it was originally permitted to have.

Still, any help was a good thing. "There were two buckets of funds," explains Taylor. "One had $92 million for cleanup and the other pot $23 million for restoration of natural resources, which would be administered by the Navassa Trustee Council."

The public then had a chance to offer written proposals for how to spend those funds. More than 20 were submitted. Many addressed river pollution problems and came from conservationists who were predominantly white and lived outside of Navassa. In contrast, there was almost no input from the town's mostly Black residents who lived near the most polluted areas.[6]

This was an environmental justice problem. "There was no intentional discrimination," Taylor points out. "That's so rare anymore, and we weren't expecting any." Instead, the difficulty was that the local conservation community wasn't racially and ethnically diverse and consisted of several organizations always on the lookout for a chance to submit federal applications. What happened, she says, was a "perfect example of some people not being in the room. It was not clear to the community that anyone could have submitted proposals." Nor was it clear to most residents what kinds of projects met the requirements.

That's when Taylor and the SELC stepped in and convinced the board to hold public meetings in Navassa to get feedback from any stakeholders in the matter. These sessions took place in churches right after Bible classes ended, which gave Navassa residents, many who suffered from creosote pollution, a chance to speak. Some suggested that any funds should stay in the community and not be spent on restoration projects far from home. Others were unhappy that the superfund site itself blocked entry to the river they once enjoyed and asked that this be remedied. Education and job training were on the wish list, too. The citizens of Navassa also wanted to create a website to acknowledge the history and cultural heritage of their ancestors, the enslaved West Africans who once populated their region—the now federally recognized Gullah-Geechee Corridor running from Jacksonville, North Carolina, to Jacksonville, Florida. Hearing these comments, observes Taylor, helped outsiders understand that the damage done by the creosote plant to residents was far worse than many had imagined.

In 2019, the federal government proposed a first round of proposals. Out of the $10 million provided for natural resources and restoration, more than $6 million was set aside for projects that benefited the community. This decision pleased Taylor, but she also worried that not enough was being done to assist workers who suffered personal injuries on the job at the creosote plant and had often worked with hazardous material without appropriate clothing, gear, and safeguards.

Today, cleanup, restoration, and remediation projects continue. As with so many environmental justice problems, there is plenty of work to do.

You Know It Won't Be Easy

Achieving environmental justice is tough, but having a mighty federal agency as an ally should be a big help. This was hardly the case in Louisiana on August 22, 2024, when a federal court struck down the EPA's effort to connect a pollution case with civil rights law. The agency previously had invoked Title VI of the federal Civil Rights Act of 1964 to withhold federal funds from any entity discriminating against others based on race or national origin. And this was happening in Louisiana, argued the EPA, where underregulated fossil fuel companies failed to decrease carcinogens from spewing disproportionately upon African Americans living in contaminated areas.

Wrong, ruled the presiding judge. It's not enough to prove outcomes weren't fair; federal civil rights law obliges you to demonstrate that what polluters did was the result of a racial-biased intent to discriminate.

Polluters and their allies were exuberant. The federal government had its intrusive knuckles rapped! Others, though, especially the mostly Black victims in Cancer Alley, were devastated. The factory workers and their communities once again were left empty-handed to work and live near acrid, toxic, smoke-belching factories. History wasn't repeating itself, but the judge's decision had a familiar sound. Didn't it resemble what plaintiffs in Houston heard in 1979 when their lawsuit to halt construction of a new landfill in Houston failed because they hadn't proven discriminatory purpose?

American law, of course, promises a fair process, not a just outcome. Achieving climate justice in court is especially tricky for the moment. Climate law is a relatively new legal area, so its repository of legal material that can be used to support a case is scant compared with other bodies of law that have had scores of years, if not centuries to develop a web of laws, decrees, acts, and policies, resulting from legislation, executive actions, and court reviews. Until quite recently lawmakers have been unsure, uninspired, or even remiss in passing legislation to cut emissions, boost alternative energy sources, fund mitigation and resiliency projects, and resolve climate disputes such as those dealing with social justice and holding accountable those who pollute the atmosphere with carbon or lie and mislead the public about the threat of climate change.

Today, the situation is changing. Climate law is growing as people around the world grapple with the legal dimensions of climate-related activities associated with mitigation, adaptation, and carbon emission reduction. Fueling much of this expansion is a noticeable uptick in climate litigation, as an increasing number of helping plaintiffs have gone to court seeking compensation from governments, corporations, and individuals that failed to protect property, human rights, and the natural environment from climate change. In 2015 Norton Rose Fulbright, a global law firm, reported over 1,400 climate cases had made their way into court in 33 countries. By July 2025, that figure had reached 3,099. The variety of climate litigations, the firm noted, has also increased.

And many of these cases are commanding the attention of the world's top jurists. "Climate litigation, including those challenging fossil fuel projects, is more often reaching the highest courts around the world," according to a June 25, 2025, analysis by the Grantham Research Institute on Climate Change and the Environment at the London School of Economics and Political Science. Over 276 climate-related cases, concluded the authors of the analysis, reached "apex courts—such as supreme courts and constitutional courts—across the world since 2015, with the United States accounting for the largest share."[7]

One of the first of these cases concerned human rights violations. The plaintiff was Ioane Teitiota, a resident of the Pacific-island country of Kiribati. In 2007, he, along with his wife and family, moved to New Zealand to start a new life. Three years later, however, the New Zealand government ordered the family to leave the country because their visas had expired. Teitiota appealed the order, arguing the lives of his family were at risk if they returned to his native island because it was being swamped by sea level rise. Thus, he argued they were climate refugees and by ordering them to return home the court would put them all at risk, a violation of international law.

After New Zealand's Court of Appeal and Supreme Court accepted the case and denied his appeal, Teitiota filed a complaint with the UN's Human Rights Committee, which also ruled against him, declaring his family was not at immediate risk. However, the committee did note that its decision "did not mean that environmental degradation resulting from climate change or other natural disasters could never create a pathway into the Refugee Convention or protected person jurisdiction." Thus, the court kept the door open for future climate civil cases.

In 2015, another climate-related case appeared in the Netherlands—a huge one. This time, the Dutch Urgenda Foundation—an NGO that promotes societal sustainability—and 886 Dutch citizens argued that the national government was endangering human life by not taking steps to reduce carbon emissions to a safe level. The court agreed with the plaintiffs and ordered the Dutch government to reduce GHG emissions at least 25 percent by 2020; four years later its decision was upheld by the Netherlands Supreme Court. This landmark decision raised public awareness about climate change and inspired similar litigation worldwide.

Although judicial decisions made in foreign countries have no standing in American courtrooms, they often influence the thinking of those involved in the legal process. US judges, for instance, may be keen to know what foreign jurists think about a legal matter, if there is no clear legal precedent for the case before them. Among the reasons the US Supreme Court noted in its majority opinion, for example, that the US Constitution forbids the execution of mentally impaired murderers, and a Texas ban on homosexual sex, were perspectives from European courts concerning similar cases.[8]

Many attorneys also keep up with climate-change legal cases in foreign countries, as they are eager to learn new legal arguments, counter arguments, and strategies. In 2024, researchers at the Columbia School of Law spotted several novel civil action themes emerging around the world. They noticed, for instance, an uptick in cases in which litigators were attempting to prove that polluters were liable for climate harm and climate-washing efforts and were misleading the public about the risk of climate change. There were also more legal complaints against governments bodies for failing to act responsibly in combating climate change, providing citizens with healthy environments, and recognizing legal protection of people displaced by climate disruptions.[9]

In December 2023, one of these new types of cases appeared in the United States. The Jean Charles Choctaw Nation submitted a formal federal civil rights complaint against the State of Louisiana for alleged discriminatory acts taken against them during a federally funded resettlement effort to relocate these Indigenous people away from their ancestral land on a narrow strip of land—Isle de Jean Charles in Terrebonne Parish, Louisiana, that was being eroded by hurricanes, land subsidence from activities of fossil fuel companies, and sea level rise. The tribe, consisting of 800 members of Biloxi, Chitimacha, and Choctaw ancestry, complained that state authorities had ignored their sovereignty, failed to properly consult with them during the relocation process, and provided substandard housing in a flood zone as a replacement. The relief the people of Jean Charles Choctaw Nation wanted was to restart negotiations with the state, which they hoped would give them greater control over their own destinies.

In recent years, several young activists have also become climate litigants. In February 2022, 13 student climate activists ranging in age from 10 to 19 filed a suit against their home state of Virginia, alleging their constitutional and public trust had been violated because state officials issued permits for fossil fuel extractions inside the state that resulted in air pollution and greenhouse gas emissions that exacerbated climate change and hurt the health of the plaintiffs. The plaintiffs also sought to convince the court to declare parts of the Virginia Gas and Oil Act unconstitutional.

Their arguments, however, failed to impress Richmond Circuit Court Judge Clarence Jenkins Jr., who presided over the case. In September of that year, he dismissed the suit, finding the state was protected by sovereign immunity.

There was much better news for activists, however, in August 2023 when a Montana state court *agreed* with a slate of young litigants who argued the state constitution's guarantee of a right to a healthy environment had been violated when state authorities tried to exempt consideration of greenhouse gas emissions in mandatory state environmental assessments.[10]

Next, on December 12, 2024, six young residents of Miami-Dade County, Florida, also went to court. They sought a declaratory judgment of their claim that the Florida Public Service decision to approve a utility's 10-year electricity plan had "locked in for decades a fossil fuel-dependent electricity sector in Florida that emits nationally and globally significant amounts of climate pollution into the atmosphere . . . [and] has prevented the promotion of clean renewable energy that the Legislature said was necessary since 2006. The Commission's conduct is knowingly worsening already dangerous conditions in Miami-Dade County that impair plaintiffs' lives and their prospects for a livable future in Florida."

Climate change, the plaintiffs also pointed out, "is already harming Miami-Dade County's critical infrastructure and unique natural resources to the point where the county is widely considered 'ground zero for climate change' in Florida and the United States."[11]

A final decision on the case was expected in early 2026.

Governments Sue Too

In recent years, a different breed of climate litigants has been showing up in the courtroom: governments, both big and small. As of early 2025, Annapolis, New York, Baltimore, San Francisco, and 10 other cities had also sued Big Oil, along with 13 counties and two tribes. States and territories involved in lawsuits against fossil fuel companies include California, Connecticut, Delaware, District of Columbia, Maine, Massachusetts, Minnesota, New Jersey, Puerto Rico, Rhode Island, and Vermont.[12]

On September 9, 2020, Charleston, South Carolina, became the first municipality in the American South to take climate legal action when it sued 24 fossil fuel companies, including BP, Exxon, ConocoPhillips, Chevron, and Shell, alleging these companies "have known for nearly half a century that unrestricted production and use of fossil fuel products create gas pollution that warms the planet and changes our climate. . . . They have nevertheless engaged in a coordinated, multi-front effort to conceal and deny their own knowledge of those threats, discredit the growing body of publicly available scientific evidence, and persistently create doubt in the minds of customers, consumers, regulators, the media, journalists, teachers, and the public about the reality and consequences of the impacts of their fossil fuel pollution." The relief sought by Charleston was a jury trial in which the city would seek relief for compensatory and punitive damages.[13]

In September 2025, it became clear that such relief was unlikely to appear any time soon, after a South Carolina judge struck down the Charleston lawsuit, inflicting a serious blow to the city's effort to hold Big Oil accountable. The city, under new leadership, decided not to appeal.

Charleston wasn't the only one to face blowback. By now there was stiff pushback from the fossil fuel industry, Alabama, 18 other GOP-led states, and the Trump administration that had been filing their own lawsuits meant to thwart both climate change legislation and litigation. Thus, climate litigation had become a two-sided affair.

Unlike many other civil lawsuits, an environmental case must provide compelling science-based evidence for the plaintiff to win. And that's not easy. It may be true that a plaintiff has brain cancer, for example, but convincing a jury with arguments backed by unassailable scientific proof that the tumor was caused by a nearby overhead high voltage power cable and not some other reason is never easy. It is even harder, for the moment, to convince a jury or a judge that a particular extreme weather event was made worse by climate change.

This could change, say legal experts, once climate science is advanced enough to demonstrate such linkage is considered "robust" and admissible by the courts. That day could be moving closer all the time, thanks to progress being made in the field of *attributive science*—a new area of study that investigates the possible links between climate change and weather patterns.

Rights of Other Living Things

And if attributive science does make those grand strides, could environmental attorneys one day use new scientific insights to bolster the assertions that just like humans, all forms of nature, whether flora, fauna, water bodies, or even entire ecosystems, also have a right to exist?

The question isn't frivolous. Many Indigenous people have long held that humans are not the master of nature but rather part of the web of life, of which all things have the right to exist. Similar thinking comes from proponents of Deep Ecology—a branch of environmental philosophy that asserts the well-being of earth's nonhuman life forms, even ecosystems, should not depend on whether humans think about them.

Legal theorists have been debating this issue, too, ever since Christopher Stone, an environmental scholar at the University of California law school, published his 1972 ground-breaking article: "Should Trees Have Standing? Toward Legal Rights for Natural Objects." Stone wasn't being facetious. "I am quite seriously proposing," he wrote, "that we give legal rights to forests, oceans, rivers and other so-called 'natural objects' in the environment."[14]

In 2001, US cultural historian Thomas Berry took up the issue by introducing a moral foundation for an idea called Earth Jurisprudence. The universe, he argued, and not humans, is the ultimate source of sovereignty, one that bestows rights to all living things to exist and to evolve.

Five years later, the local council of Tamaqua Borough in the coal region of Pennsylvania validated this idea when it became the first community on earth to recognize legal rights for ecosystems. This gave the local council or residents on behalf of a local ecosystem the ability to take legal action to stop companies from polluting the community. Ben Price, the Community Environmental Legal Defense Fund's (CELDF) national organizing director, collaborated closely with the community's leaders to get that recognition. Striking a blow against climate change, however, he points out, was not the main goal of the residents of Tamaqua. "A little community doesn't think about everyone else," says Price. "They are angry they were treated as a sacrifice area. And they will do what they can to fight back."[15]

Thus, the city's ordinance was meant to offer residents a handy bludgeon to protect themselves. "I guess we adopted that idea of rights of nature and waited to see what happens."[16]

What happened is that they have won, at least until all inevitable legal challenges are settled. No matter what the outcome may be, the extraordinary action taken by the community of Tamaqua signaled that the winds of change were blowing. The community also inspired others. Today, rights of nature laws appear in constitutions, treaty agreements, statutes, local ordinances, and court decisions in 17 countries.

In the American South, the idea took root in several places, including Austin, Texas; Asheville, North Carolina; and Buckingham County, Virginia. In November 2020, Orange County, Florida, voters embraced the idea of "wild law" when an astounding 89 percent of local voters approved amending their city charter—which in Florida means a kind of constitution that grants the community greater self-rule power than it would have under state law. The amendment required the well-being of natural systems be taken into consideration by officials when making development decisions. It was the first such action in the nation. Elsewhere in Florida, the cities of Titusville, Venice, Fort Myers, and Naples took similar actions.

Before the election took place, however, a lawmaker in the Republican-dominated state legislature with ties to Big Agriculture slipped a few lines into a 111-page omnibus bill that forbade local governments from granting special rights to nature as the voters of Orange County had in mind. The measure generated little publicity at the time, and so when the November election arrived, few Orange County voters realized their state government had already quietly outlawed their nature-protecting ordinance. Many were outraged. Once again Florida lawmakers had thwarted the efforts of citizens to protect the natural world. Soon, a test case was filed, invoking the new Rights of Nature measure to stop a developer whose construction plans the plaintiffs believed would harm water bodies in Orange County. In July 2022, a Florida circuit court judge declared state law trumps a local amendment, and the Orange County rights of nature issue fizzled out.

The story, however, was not over. Not long after the amendment was canceled, another group, inspired by the Orange County initiative, began a campaign of its own to put a measure on the state ballot for voter approval of a proposed amendment to the state constitution. If ever approved, it would create a citizen's fundamental right to clean and healthy waters. That could be an important tool in protecting freshwater reserves now under threat from sea level.

But is a rights of nature law the best way to proceed? Today, in retrospect, CELDF's Ben Price thinks, "Arguing nature should gain status as a legal person is a mistake." He worries that once society agrees a river has rights, "the question then is can you hold it responsible for damages like when it floods?"[17]

A better way, Price suggests, is for a human to serve as a *legal guardian* for a river, animal, or ecosystem in court. The Guardian Ad Litem system works this way when caring adults protect the best interests of children in court.

In 2017, Price took this idea to Buckingham County, Virginia, where hundreds of grassroots volunteers had teamed up to oppose a proposed Atlantic Coast Pipeline (ACP), a 600-mile fracked-gas pipeline slated to run through the state. Worse yet, the plans also called for a super-sized compressor station to be installed in Union Hill, a mostly African American community located in the county, On July 5, 2020, after battling the activists for six years, ACP officials admitted defeat, claiming the price tag for the pipeline project was too dear.

The Buckingham activists weren't done, however; they now wanted to be ready for the next fight. That's why they created the Virginia Community Rights Network in partnership with Ben Price to empower themselves to act as guardians when the time came "to elevate the rights of people and Nature over illegitimate corporate rights that impede local and direct democracy."[18]

Today, like so many other green initiatives, this movement has faced fierce opposition from the time Donald Trump returned to the US presidency in January 2025. From the onset, he and his minions were quick to wreak havoc on policies and norms they had long despised. And they seemed to take special delight in shredding hard-earned climate actions across the nation.

THIRTEEN
THE POINT OF NO RETURN?

On the night of December 7, 2022, an African American pastor delivered a soaring victory speech in an Atlanta hotel ballroom to a jubilant mixed-race audience. "In a sense, I am Georgia!" he exclaimed. "A living example and embodiment of its history and its hope, of its pain and promise, the brutality and possibility."[1]

These words were historic and everyone in the ballroom knew why. Savannah-born Reverend Raphael Warnock had just won a seat in the US Senate in a hotly contested statewide runoff election. Not only that, two years earlier, he had also won a special election and earned the right to finish another senator's term of office. That meant Warnock became the first African American Democrat to represent Georgia in the US Senate, and also the first from any southern state which once belonged to the Confederacy.

Warnock had more to say: "And Georgia, once again—as you did in 2021, when you sent an African American man and . . . [Jon Ossoff, the first Jewish senator from Georgia,] in one fell swoop—you are sending a clear message to the country about the kind of world we want for our children."[2]

At the time, Warnock's victory seemed to signify something important had changed in Georgia, but perhaps in the entire South, as well. In 2018, Bob Moser, the great modern chronicler of Southern politics and self-professed optimistic progressive, thought he saw it coming when he wrote, "Georgia today is a microcosm of the changes that have radically remade the South since the civil-rights revolution." Since the 1970s, when big banks and tech companies and foreign-based corporations started moving their headquarters to cities like Charlotte, Atlanta, and Houston, the region has basically been one massive construction site. Professional whites, remigrating Black people, and the nation's biggest share of Latino and Asian immigrants, he adds, means "the old insular South was changing fast."[3]

So much African American reimmigration had been underway that pundits referred to it as a "New Great Migration," which started in the 1990s, when the net number of African Americans moving to the South exceeded the number leaving.[4] At the head of the southbound parade were college-educated adults dubbed a "brain gain" that helped Warnock win.[5] Like everyone else, incoming Black Americans are seeking economic improvement, better schools, and safe neighborhoods; they are also hoping the Old South their grandparents fled is expiring.

There's also another significant trend. It's the growing number of Asian Americans, who according to the US Census Bureau are the fastest-growing ethnic group in the nation and in the South.[6] According to a 2022 collaborative study conducted by UNC-Chapel Hill's Asian American Center and Carolina Demography between 2010 and 2020, the AAPI population across the Southeast increased by 46 percent. In North Carolina that figure soared to 64 percent, boosting AAPI to 4.1 percent of the state's population.[7]

These new arrivals, along with all the other newcomers, have delivered a powerful economic clout to the region. During 2020 and 2021, in fact, they generated $100 billion in economic activity in the Southeast, while the Northeast lost about $60 billion, according to a Bloomberg analysis of Internal Revenue Service data.

A massive influx of businesses, skilled workers, and investments—why, the very things Henry Grady and the Southern oligarchs had desired so fervently so long ago—were now arriving. As Margaret Renkl, the Tennessee essayist, noted, "Georgia is the clearest proof yet that this is not our grandfather's Southland. And it will never be again."[8]

"Yes, the Old South is dead," agrees Alex Easdale, executive director of the Southeast Climate and Energy Network, a donor-based organization that helps local organizations develop equitable and sustainable energy systems across the Southeast. Despite all the South's ongoing problems, Easdale looks at the region through the "lens of opportunity." "It's one of the most dynamic regions in the world, extremely diverse, very vibrant, a lot of culture and history," he effuses. "No, this isn't the Old South."[9]

A big question, however, nags Southerners who read the latest bulletins from the IPCC. Will these new residents help invigorate a robust reaction to climate change and create a more just, equitable society? Or will they fuel consumption, consume lots of energy, clear forests for new housing and commercial buildings, and thus dump more GHG into the atmosphere?

And what will happen if conservatives view all these new faces as further proof they are indeed being replaced, just as far-right white nationalists had been warning? If so, will they harden their resolve and buck even harder against perceived "progressive" goals such as climate action? After all, what was happening in Florida provided a hint of reactionary moves yet to emerge elsewhere in the South.

Florida Heads Backward?

The Sunshine State, in fact, has also been exploding with population growth for years, but its trajectory is far different from what progressives had hoped for. You might even say the state was becoming "Dixie-fied" again. At least that's how it appears to author, journalist, and commentator Diane Roberts, a professor at Florida State University and fellow in creative writing at the University of Northumbria in England, where she specializes in Southern United States culture. "Florida was once a New South state," she opined in a November 2022 article for the *Florida Phoenix*, an affiliate of States Newsroom, the nation's largest state-focused nonprofit news organization. "From the years 1960s to 2000 or so, Florida had leaders who looked at the future, determined to leave Jim Crow behind and separate ourselves from the likes of Alabama. We had governors committed to equal justice, open government and voting rights—however imperfectly achieved."[10]

Those days were now long gone with Ron DeSantis as governor, she insisted; "Florida is sliding back into the mire of its Old South past." In Roberts's view, the suspected retrogression is the result of a rash of legislative actions taken in Tallahassee under the governor's watch, such as gerrymandering of congressional districts, the creation of Election Police, and passage of laws that critics allege prevent honest discussions about racial matters in university classes.[11] Florida's Department of Education also caused a national uproar when it issued new school curriculum guidelines about the teaching of slavery that many critics view as an overt attempt to sanitize an ugly racist past.

During his 2022 reelection campaign, Florida's governor, however, proudly ran on these same achievements and won with 60 percent of the statewide vote. Neither his blatant climate change denialism, nor his heavy-handedness in the culture war appeared to weaken his political power.

The World Keeps Warming

Meanwhile, troubling climate news kept piling up. Add it all up and you would have to conclude those IPCC climatologists had been wrong after all—but not for any of the reasons touted by denialists. That's because many of those anticipated extreme events seemed to be arriving way *ahead* of schedule. "The climate crisis has arrived and is accelerating faster than most scientists expected," warned an independent assembly of researchers, the Alliance of World Scientists, in 2020. "It is more severe than anticipated, threatening natural ecosystems and the fate of humanity."[12]

The bad news kept coming. In 2023, Kentucky, for instance, underwent a state of emergency amid unprecedented flooding. Uncontrolled fires burned Canada and Greece. In Rome, temperatures hit 107°F, and a record-setting 126°F in China. Death Valley baked in 128°F heat, while in Phoenix the daily average temperature for more than a week was 108°F. The worst-hit place on earth was the Middle East, where thermometers registered 152°F, pushing human life to the brink of survival. United Nations Secretary-General António Guterres predicted that what was happening that summer of 2023 would likely "shatter records across the board."[13]

His message was grim. "Climate change is here," Guterres said. "It is terrifying. And it is just the beginning." He sounded like a visitor from the Apocalypse describing an alien world where he saw "Children swept away by monsoon rains, families running from the flames [and] workers collapsing in scorching heat."[14]

Other international bodies also sounded the alarm. According to the World Meteorological Organization and the European Commission's Copernicus Climate Change Service, during July 2023 the planet experienced both the hottest three-week period and the hottest three days ever recorded. In fact, at the time, the whole year had been the most scorching since humans began keeping track of temperatures.

The American South escaped none of this misery. Extreme heat combined with tropical moisture made it feel like 105 or even 110°F in some Gulf Coast states. Ocean water off the Florida coastline was a shocking five degrees above normal. The temperature in Manatee Bay in the Everglades National Park was 101°F.

By the summer of 2024, more CO_2 than ever was being released into the atmosphere, and all the climate ills associated with that phenomenon grew worse. The year, which NOAA's National Centers for Environmental Information says was the warmest year on record for the contiguous United States, saw 27 separate billion-dollar weather and climate disasters. Among them were 18 tropical cyclones in the Atlantic basin, with five making landfall as hurricanes, Hurricane Helene heading the list as most destructive.

A year later, a massive heat dome spread out across great swaths of the United States, causing 220 million American to swelter in temperatures that often topped 100°F. Conditions for many people in the early summer, meteorologists warned, were "dangerous, life-threatening."[15]

Great numbers of southern leaders seemed unbothered by what was happening in real life. Instead, many went out of their way to aggravate the situation. As many communities sweltered in triple-digit temperatures of 2024, for instance, Texas state lawmakers ended city and county ordinances in progressive outlier cities, such as Dallas and Austin, requiring 10-minute breaks every four hours for outdoor workers. Union leaders expressed alarm that lawmakers supported such a measure in a blistering state that already led the nation in worker deaths from extreme environmental heat.

In that same year, Texas legislators, along with their counterparts in North Carolina and Florida, were also trying to limit what teachers and students could say about climate change in public school classrooms.

Lost at the Crossroads?

Such obstruction doesn't change the reality that everyone now stands at the crossroads in the era of climate change, and nobody can dodge having to make big decisions about how to react to what is clearly happening. In March 2022, Cynthia Tucker, a Pulitzer Prize–winning journalist and native Alabamian who coauthored *The Southernization of America: A Story of Democracy in the Balance*, told National Public Radio's Debbie Elliott, "I know that it is up to everyday Americans to decide which way we are going to go in this country. It would be easy for people to follow the political leaders who are demagoguing on fear and resentment. That is the easy path. But it is also quite possible for everyday Americans to turn away from that and to say, no, I understand that America can continue to be a great, multiracial, pluralistic democracy, and that's the way we want to go."[16]

On November 5, 2024, a reply to Tucker's musing appeared when a majority of voters once more elected Donald Trump as their next president; by now he was a convicted felon who had faced scores of other indictments. Still a hard-core climate change denialist, he soon withdrew the United States from the World Health Organization and the Paris Climate Accord, and he scrapped a significant portion of Biden's green and justice initiatives, while boosting support for the extraction of fossil fuels. He also placed a climate denier at the helm of the EPA, an agency hit with devastating budget cuts. (Almost three years earlier, the US Supreme Court had restricted EPA's ability to regulate GHG emissions—a boon to polluters.) Under Trump, NOAA announced it would no longer monitor the cost of climate change–fueled weather disasters, despite the fact that scientists depend on such reports to assess future extreme weather events. An executive order from Trump directed the Department of Justice to challenge state climate action laws. Trump also moved to halt construction of five offshore wind projects off the coasts of Virginia, New York, and New England.

As Trump shocked the nation and the world with additional unprecedented, and possibly unconstitutional, exertions of presidential power abroad and at home, seemingly steeped in grievance and revenge, the GOP tried to normalize and whitewash his behavior. Factions in the party also propagated a new Lost Cause Myth—namely that Trump hadn't lost the 2020 election, hadn't caused an insurrection, and hadn't tried to prevent an orderly transition of presidential power. Rather, he was persecuted by a mysterious deep state cabal and unjust legal systems. Just like its Southern predecessor, the new myth seeks to redeem those who took part in insurrection against the United States and recast them as patriots.

Trump would not be back in the White House without southern backing. All southern states—Georgia included—voted for the man who called climate change a hoax.

No matter what their reasons were for supplying Trump this assistance, these supporters inarguably helped steer a disastrous turn at the climate crossroads that will make the approaching climate challenges all that much worse.

An Uncertain Future

The day after the 2024 US presidential election, Laurence Tubiana, a lead negotiator of the 2015 Paris Climate Agreement, rang a warning on the social media platform X: "To be clear: the US election result is a blow in the fight against the climate crisis. The window to limit warming to 1.5°C is closing—these next 4 years are critical. But let's not despair. The Paris Agreement has proven resilient, stronger than any one country's policies."[17]

But can it succeed without the help of the world's most powerful and influential country? It's hard to understand how this could be possible. However, even with US leadership, the world is not organized to deal with global challenges like climate change, Stefan Lehne, Senior Fellow at Carnegie Europe told this writer. "We would need a strong central institution with sanctioning powers. Instead, we have the 'National Determined Contributions' of the Paris agreement without any meaningful enforcement mechanism." Therefore, Lehne explains, "most of the mobilization for action has to happen on the national level."[18]

If so, this means states must be the ones to "take the lead, both in terms of actions and shaping of public consciousness," argues Anatol Lieven in his book *Climate Change and the Nation State.* They have the power of the purse, authority, resources, manpower, and political heft to make a big difference. And these same nation states, Lieven adds, will be "pushing against immense gravitational forces; not just resistance to the radical economic reforms that will be required, and the power of vested interests, but against contemporary culture itself, with its intense materialistic individualism and shorter attention spans."[19]

All the same, Anatole Lieven thinks nationalism with the right kind of leadership could be harnessed to galvanize public participation in addressing climate change. What nations must do, he argues, is seize the power and energy of growing nationalism within their own country and repurpose it as a fight against the external enemy known as climate change, at least for security reasons. "To defeat the Trumps and Bolsonaros [the former president of Brazil] on climate change," he argues, "it is necessary to seize the weapon of nationalism and use it against them."[20] Moreover, he adds, "What especially brings nationalism and ecological thinking together is the capacity for both to demand sacrifices on the part of existing people for the sake of future generations."[21]

A sparkling example of this kind of nationalism is on display in Uruguay, a small nation in South America which in 2008 began weaning its electric grid off fossil fuels. As of 2024, the nation's electric power was 99.1 percent composed of alternative renewable sources, lots of it from an abundant wind supply, along with a mix of solar, biomass, and hydropower resources. Leading his nation's energy transformation was Ramón Méndez Galain, a particle physicist—an analytical person, not an ideologue. Today, private companies own and run the wind turbines, and the state buys the energy and distributes it. Energy is cheap, too. Better yet, Galain's plan may have created 50,000 new jobs.[22]

If Uruguay with a population of 3.4 million can succeed, why can't any southern state, each large enough to be a nation, do the same? What may be needed is a strong dose of Galain's scientific expertise, pragmatism, political will, and continuity of governance to get the job done.

The alternative of continuing to heat the planet on purpose, or retreating or sticking your head in the sand, won't help. "All too many people fail to remain awake through great periods of social change," Dr. Martin Luther King Jr. wrote. "Every society has its protectors of the status quo and its fraternities of the indifferent who are notorious for sleeping through revolutions. But today our very survival depends on our ability to stay awake, to adjust to new ideas, to remain vigilant and to face the challenge of change. The large house in which we live demands that we transform this worldwide neighborhood . . . or together we will be forced to perish as fools."[23]

Everybody in that large house must pitch in. As IPCC Working Group II Cochair Debra Roberts puts it, "Our assessment clearly shows that tackling all these different challenges involves everyone—governments, the private sector, civil society—working together to prioritize risk reduction, as well as equity and justice, in decision-making and investment."[24]

There's also plenty that everyone can do to help at the *individual* level, especially while waiting for responsible leadership to arrive. Activists offer a few suggestions: Reduce personal carbon footprints by eating more vegetables and consuming less meat and dairy products. Drive less and walk, and travel by bicycles and public transportation more. Fly less, too. Insulate homes, businesses, and public places. Avoid plastic. Turn off electric appliances not being used. Wash clothes in cold water. Reduce waste and recycle. Read lots about climate change and don't spread misinformation. Remember life is unfair and many vulnerable people may need help. Above all, put pressure on elected officials to implement reform measures to ward off the worst of climate disorder. Vote only for those who understand this.

Finally: Don't despair. The current climate action backlash won't last forever, even in the conservative South. In fact, as the nation's most populated, most conflicted, and perhaps most intriguing region continues to redefine itself, you can only wonder if a new kind of Southernization is now shaping the lives of younger Southerners who are appalled by their elders' refusal to wake up to danger.

As always, change is inevitable. Legions of climate activists remain optimistic that the South will be able one day to escape the shroud of denialism and manage any upheavals that come its way, at least as well as any other region of the country. After all, Southerners have had lots of practice coping with big troubles in the past.

So, is Cynthia Tucker right? Will the South someday soon rise to face an existential threat and this time prevail in a cause worth winning? Or will the debilitating impact of denialism, indifference, selfish interest, and tribalism hasten the coming of a point of no return for a collapsing natural world? The planet is going to find out soon enough, because the seas keep rising. Just look at Fort Sumter in Charleston Harbor. It's going under.

NOTES

Introduction

1 IPCC, 2023: Summary for Policymakers. In: H. Lee and J. Romero, eds. Climate Change 2023: Synthesis Report, Contribution of Working Groups I, II and III to the Sixth Assessment Report of the Intergovernmental Panel on Climate Change, IPCC, Geneva, pp. 1–34. https://doi.org/10.59327/IPCC/AR6-9789291691647.001.
2 Caitlin Rosenthal, "Plantations Practiced Modern Management," *Harvard Business Review,* September 2013. https://hbr.org/2013/09/plantations-practiced-modern-management.
3 Silkenat, *Scars on the Land,* 202.
4 Grady, "The New South."
5 "Restoring the Quality," 126.
6 National Advisory Commission.
7 National Advisory Commission, 5.
8 Maloney, "African Americans."
9 Maloney, "African Americans."
10 Daniel, *Standing at the Crossroads.*
11 Kromm, "As the South Goes."

One Hot Time in the Southland

1 Meyer, "American South."
2 Pierre-Louis, "If You Live in the South."
3 Solomon et al., "Estimating Economic Damage."
4 Cusick and *E&E News,* "This Part of the U.S."

5 Schleifstein, “Climate Change.”
6 “What Are Coastal Wetlands?” Environmental Protection Agency. https://www.epa.gov/wetlands/about-coastal-wetlands.
7 “For Wetland Plants.”
8 L. Donelson Wright, former director of Marine Science at William & Mary’s Virginia Institute of Marine Science, interviewed by John M. Dunn, February 2, 2023.
9 Wright interview.
10 Wright interview.
11 Nicoletti, “Mangroves Need Everglades Restoration.”
12 Yates, “Rising Temperatures.”
13 Howard, “Best and Worst States.”
14 First Street Foundation, “3rd National Risk Assessment,” 4.
15 Margie Vicknair-Pray, Sierra Club Delta Chapter conservation program coordinator, interviewed by John M. Dunn, October 24, 2022.
16 Vicknair-Pray interview.
17 Karst, “Mid-Barataria Sediment Diversion.”
18 Vicknair-Pray interview.
19 Matt Hauer, assistant professor of sociology at Florida State University, climate migration specialist, interviewed by John M. Dunn, October 26, 2022.
20 Hauer interview.
21 Berlin Rubin and Wong-Parodi, “As California Burns,” 15–45.
22 North American Moving Services, “2021 Migration Report.”
23 “Climate Justice,” *Sustainable Development Goals, United Nations,* May 31, 2019. https://www.un.org/sustainabledevelopment/blog/2019/05/climate-justice/.
24 Hauer interview.

Two From the South a New Civil Rights Movement

1 Helman, “America’s 20 Worst.”
2 Falcon and Nexstar Media Wire, “10 Most Polluted States.”
3 Mayhew Bergman, “They Chose Us.”
4 Mayhew Bergman, “They Chose Us.”
5 “Study Finds Exposure to Air Pollution.”
6 “Latinos Communities.”
7 “Come Hell or High Water.”
8 “Come Hell or High Water.”

9 "10th Anniversary of Hurricane Katrina."
10 Tuckey, "Feds OK."
11 "EPA Starts a Civil Rights Probe."
12 Williams, "'A Great Day.'"
13 Horning, "Social Network."
14 Pittman, "FL Sacrificed a River."
15 Horning, "Social Network," 40.
16 Horning, "Social Network," 44.
17 Dearen et al., "AP Finds Climate Change Risk."
18 *City of Houston v. George*. 479 S.W.2d 257 (1972). https://law.justia.com/cases/texas/supreme-court/1972/b-2726-0.html.
19 Watkins, "Houston's 'Father.'"
20 Watkins, "Houston's 'Father.'"
21 Watkins, "Houston's 'Father.'"
22 Bullard, "Environmental Justice for All."
23 Bean v. Southwestern Waste Management Corp., 482 F. Supp. 673 – Dist. Court, SD Texas (1979) https://law.justia.com/cases/federal/district-courts/FSupp/482/673/2095959/.
24 *Bean*, 482 F. Supp. 673.
25 Burwell and Cole, "Environmental Justice."
26 United States v. Robert Earl Ward, Jr. A/k/a "buck" Ward, 676 F.2d 94 (4th Cir. 1982). https://law.justia.com/cases/federal/appellate-courts/F2/676/94/375500/.
27 Labalme, "From the Archives."
28 Burwell and Cole, "Environmental Justice."
29 Brodkin, *Environmental Activism*, 50.
30 Katz, "Birth of an Environmental Movement."
31 Mayhew Bergman, "They Chose Us."
32 Labalme, "From the Archives."
33 McGurty, "From NIMBY to Civil Rights," 301–323.
34 Summers, "PCB Protest."
35 "Afton, NC (Warren County)."

Three The Movement Grows

1 GAO, "Siting of Hazardous Waste Landfills."
2 "Toxic Waste and Race."

3 Heritage, "From the Editorial Page," 2.
4 Bryant and Mohai, "Michigan Conference," 9.
5 "History of Environmental Justice."
6 Bullard, "Environmental Justice."
7 Middleton, "Battle of Carver Terrace."
8 "People of Color."
9 "People of Color."
10 Alston, "The Summit."
11 Alston, "The Summit."
12 Alston, "The Summit."
13 Chang and Hwang, "It's A Survival Issue."
14 Bullard, "Environmental Justice."
15 "Who Is APEN?"
16 Choi, "Why the Environmental Movement."
17 United Nations Framework Convention on Climate Change, 1992. https://unfccc.int/resource/docs/convkp/conveng.pdf, 4.
18 *We Choose Now.*
19 *We Choose Now.*
20 Whitehead, "First Climate Justice Summit."
21 Karliner, "Climate Justice Summit."
22 Karliner, "Climate Justice Summit."
23 Dillen, "What We Saw."

Four **Is Eco Justice Possible?**

1 "Public Good or Private Wealth?," 12.
2 Owen *et al.*, "Energy Transition Minerals."
3 Block, "Mining."
4 Jones, "Americans Say."
5 Blakemore, "Thorny History."
6 Herscher, "U.S. Government to Pay."
7 "Ten Years of Restoring."
8 Griffith, "Economist Proposes."
9 Byrnes, "West Ocala."
10 Carnel and Douglas, "Remembering Rosewood: Descendants Mark Racial Violence That Razed Florida Town 100 Years Ago," WUFT, PBS affiliate in Gainesville, Florida, January 3, 2023. https://www.wuft.org/human-interest/2023-01-03

/remembering-rosewood-descendants-mark-racial-violence-that-razed-florida-town-100-years-ago.

11 Hannah-Jones et al., *The 1619 Project, 282.*

12 “Greensboro Truth & Reconciliation.”

13 “Fact Sheet.”

Five Push Back

1 Jeffreys, “Environmental Racism.”

2 Jeffreys, “Environmental Racism.”

3 Sowell, *Quest for Cosmic Justice,* 16.

4 Sowell, *Quest for Cosmic Justice,* 13.

5 Foreman, “Clash of Purposes.”

6 Foreman, “Remarks.”

7 Tessum et al., “Inequality in Consumption.”

8 Tessum et al., “Inequality in Consumption.”

9 Taylor, “Rhetorical Appeals.”

10 Friedman, “‘Environmental Racism.’”

11 Buford, “Moment Been Lost?”

12 Bullard et al., “Toxic Wastes.”

13 Bullard et al., “Toxic Wastes.”

14 Perls, “EPA Undermines.”

15 “Biden Administration.”

16 Rothstein, *The Color of Law,* 126–132.

17 Bustamante, “I-45 Expansion Project.”

18 “Joint Letter to President Joseph R. Biden, Jr.” Republican Governor’s Association, January 19, 2002. https://www.rga.org/wp-content/uploads/2022/01/Joint-Letter-to-President-Biden-Requesting-IIJA-State-Flexibility-1-19-22.pdf.

19 “Joint Letter to President Joseph R. Biden, Jr.”

20 Dumain, “Republicans Deploy.”

21 Yuccas, “Learn the Meaning.”

22 Yuccas, “Learn the Meaning.”

23 SouthWest Organizing Project letter to Jay D. Hair.

24 Durlin, “Shot Heard Round the West.”

25 Durlin, “Shot Heard Round the West.”

26 Durlin, “Shot Heard Round the West.”

27 Durlin, “Group of 10.”

28 Jeffreys, "Environmental Racism."
29 Gardiner, "Unequal Impact."
30 "Affinity Group Letter."
31 Brune, "Pulling Down Our Monuments."
32 Brune, "Pulling Down Our Monuments."
33 Brune, "Pulling Down Our Monuments."
34 Brune, "Pulling Down Our Monuments."
35 Virginia Sanders, A National Sierra Club Environmental Justice employee and member of the Lower Richmond, South Carolina NAACP, interviewed by John M. Dunn, November 4, 2023.
36 Sanders interview.
37 Sanders interview.
38 Sanders interview.
39 Sanders interview.
40 Bullard, "Earth Month."
41 Bullard, "Earth Month."
42 Britton-Purdy, "Environmentalism."
43 Pezzullo and Sandler, "Introduction."
44 Pezzullo and Sandler, "Introduction," 5.
45 Jones, "Environmental Movement."

Six Roots of Denial

1 "2019 Climate Action Summit."
2 Bruggers, "Southern State Energy Officials."
3 "What's Wrong."
4 "Rate of Change."
5 Judith Curry, Professor Emeritus and former chair of the School of Earth and Atmospheric Sciences at the Georgia Institute of Technology, interviewed by John M. Dunn.
6 Silver, "Three Worlds."
7 Bonnie et al., "Understanding Rural Attitudes," 3.
8 Bonnie et al., "Understanding Rural Attitudes," 5.
9 Bonnie et al., "Understanding Rural Attitudes," 18.
10 Bonnie et al., "Understanding Rural Attitudes," 5.
11 "How Religious Is Your State?"
12 Smith, "The Spread of COVID-19."

13 Hofstadter, *Anti-intellectualism*, 47.
14 Hofstadter, *Anti-intellectualism*, 15.
15 Neidl, "How to Talk to People."
16 Neidl, "How to Talk to People."
17 Dickey, *Under the Eye*, 79.
18 Marjorie Taylor Greene (@mtgreenee), X, October 3, 2024, https://x.com/mtgreenee/status/1842039774359462324
19 Matt Wallace (@MattWallace888), X, September 30, 2024, https://x.com/MattWallace888/status/1840675609531359507
20 Samuels, "Biden Rips 'Beyond Ridiculous.'"
21 Grant, *Deepest South of All*, 184.
22 Grant, *Deepest South of All*, 185.
23 Grant, *Deepest South of All*, 146.
24 Benegal, "Spillover of Race."
25 Parajon, "Effect of Racial Resentment."
26 Gary Zarillo, Professor of Oceanography and Ocean Engineering, Florida Institute of Technology in Melbourne, Florida, interviewed by John M. Dunn, April 13, 2003.

Seven The GOP vs. Climate Action

1 "Lewis Powell Memo."
2 "Lewis Powell Memo."
3 Schmitt, "Myth of the Powell Memo."
4 Collomb, "Ideology of Climate Change Denial," 4.
5 Jacques et al., "Organization of Denial."
6 Shabecoff, "Reagan and Environment."
7 Shabecoff, "Reagan and Environment."
8 Bush, "President's News Conference."
9 "Chilling Story."
10 Waldman and Hulac, "When the GOP Turned Away."
11 Collomb, "Ideology of Climate Change Denial," 33.
12 McCright et al., "Political Polarization."
13 Funk and Hefferon, "U.S. Public Views."
14 "Climate Change Research."
15 Burbank, "Bush Views Shift."
16 "2008 Republican Party Platform."

17 "2008 Republican Party Platform."

18 Armey and Kibbe, *Give Us Liberty*, 90–91.

19 Maxwell, "How Southern Racism."

20 Jackson, "Southernization of American Politics."

21 John S. Jackson, political scientist and former Chancellor of Southern Illinois University, Carbondale, interviewed by John M. Dunn, May 30, 2025.

22 "2012 Republican Party Platform."

23 Plait, "Climate Denier Ted Cruz."

24 "Transcript of Donald Trump's Dec. 30 Speech in Hilton Head, S.C." Federal News Service, *Kansas City Star*, January 20, 2016. https://www.kansascity.com/news/local/news-columns-blogs/the-buzz/article55604115.html.

25 Adler and Leber, "Donald Trump."

26 Roberts, "GOP Is World's Only."

27 Lieven, *Climate Change*, 10.

28 "Republican Platform 2016."

29 "Republican Platform 2016," 22.

30 Drennen and Hardin, "Climate Deniers."

31 "Denial, Disinformation, and Doublespeak."

32 Marcetic, "Democrats Are Climate Deniers."

33 Lieven, *Climate Change*, 131.

34 Davenport and Lipton, "How G.O.P. Leaders Came to View."

Eight **You'd Better Have a Plan**

1 Irfan, "Scientists Are Divided."

2 Shepherd, "Rick Perry Says."

3 Shepherd, "Rick Perry Says."

4 Cramton, "Fostering Resiliency."

5 Cramton, "Fostering Resiliency," 2.

6 "Climate Change in Texas."

7 Douglas, "Texas' Plan to Provide Water."

8 Katherine Hayhoe with Dave Strauss, Nature Conservancy tele-townhall presentation, September 2025.

9 "Texas Coastal Resiliency Master Plan," 1.

10 "Texas Coastal Resiliency Master Plan," 46.

11 "Governor Cooper Signs."

12 Schleifstein, "Edwards' Climate Task Force."

13 Regional Greenhouse Gas Initiative. https://www.rggi.org/.

14 Ferman, "Texas Bans."

15 Alys Campaigne, Climate Initiative Leader, Southern Environmental Law Center, email exchange, February 7, 2025.

16 Campaigne, email exchange.

17 Daniel Gilford, Climate Scientist with Climate Central, interviewed by John M. Dunn, November 9, 2023.

18 Gilford interview.

19 Gilford interview.

20 "What is Climate Resilience?"

21 "James Butler and the Flood of 1927."

22 "James Butler and the Flood of 1927."

23 "James Butler and the Flood of 1927."

24 Gary Zarillo, professor of oceanography and ocean engineering, Florida Institute of Technology in Melbourne, Florida, interviewed by John M. Dunn, April 13, 2023.

25 "Exposed: Social Vulnerability," 2.

26 "U.S. Climate Vulnerability Index."

27 Chris Mooney et al., "Drowning South."

Nine Locals Lead When Those in Charge Won't

1 Ryan Freed, climate action officer for Alexandria, Virginia City Government, interviewed by John M. Dunn, November 14, 2023.

2 Gilford interview, November 9, 2023.

3 Climate Mayors Letter.

4 Steve Adler, mayor of Austin Texas, interviewed by John M. Dunn, December 12, 2022.

5 Turner, Houston Climate Action Plan, 4.

6 Zach Baumer, climate program manager, City of Austin, Texas, interviewed by John M. Dunn, October 5, 2022.

7 "Miami Residents Vote."

8 Gross and Harris, "Key Biscayne Needs to Adapt."

9 Philip Stoddard, mayor of South Miami, White House Governance Coordinating Committee of the National Ocean Council, and biology professor at Florida International University, interviewed by John M. Dunn, December 11, 2022.

10 Ritchie, "Climate Doomsayers."

11 Stoddard interview.

12 Kyle Spencer, Norfolk, Virginia, chief resiliency officer, interviewed by John M. Dunn, May 1, 2023.

13 Christine Morris, "Dutch Dialogues Recap," June 23, 2015. https://www.youtube.com/watch?v=36umjKvxtPA&t=18s.

14 Spencer interview.

15 Spencer interview.

16 Spencer interview.

17 Spencer interview.

18 Freed interview.

19 Casi Calloway, chief resilience officer for the City of Mobile Alabama, interviewed by John M. Dunn, May 25, 2023.

20 Calloway interview.

21 Calloway interview.

Ten Southern Troublemakers

1 "On Mississippi's Gulf Coast."

2 "Katherine T. Egland, National Association for the Advancement of Colored People," https://naacp.org/people/katherine-t-egland.

3 "What We Do."

4 Emilia Aguinaga, Climate Justice Program Officer, Foundation for Louisiana, interviewed by John M. Dunn, June 6, 2025.

5 "Catherine Coleman Flowers," video segment, https://www.catherinecolemanflowers.com/.

6 "Hilton Kelly," The Goldman Environmental Prize, https://www.goldmanprize.org/recipient/hilton-kelley/.

7 "Hilton Kelly."

8 "Marshall Saunders."

9 "Marshall Saunders."

10 "Marshall Saunders."

11 "Marshall Saunders."

12 Vazquez, "Climate Treason March."

13 Florida Conservation Voters.

14 Jack Sweeney, former campaign assistant with the Louisiana Bucket Brigade, interviewed by John M. Dunn, March 24, 2023.

15 Sweeney interview.

16 Sweeney interview.

17 Aaron Ambroso, codirector and cofounder of Houston Climate Justice Museum, interviewed by John M. Dunn, April 14, 2023.

18 Ambroso interview.

19 Ambroso interview.

20 Ambroso interview.

21 Myers Park Baptist Church, Environmental Justice. https://myersparkbaptist.org/serve/environmental-justice.

22 Codi Norred, program director at Georgia Interfaith Power & Light, interviewed by John M. Dunn, October 25, 2021.

23 Norred interview.

24 Norred interview.

25 Pilot Episode: *Green the Church Presents R.A.C.E.*, Reverend Dr. Ambrose Carroll. https://www.youtube.com/watch?v=eqH5E7ZqPaM.

26 Pilot Episode.

27 Jan Booher, project director on an EPA Environmental Justice Grant awarded to the Green Sanctuary Committee of the Unitarian Universalist Fellowship of Boca Raton, interviewed by John M. Dunn, May 11, 2021.

Eleven Southern-Style Green New Deals

1 Freidman, "Warning from the Garden."

2 "Green New Deal."

3 "Fact Sheet: Inflation Reduction Act."

4 Bakst, "Inflation Reduction Act."

5 "Gulf South for a Green New Deal," https://www.gulfsouth4gnd.org/_files/ugd/a491a1_637aea05b7814ad5b917daca77777118.pdf.

6 Scot Quranda, communication director, Dogwood Alliance, Asheville, North Carolina, interviewed by John M. Dunn, October 20, 2022.

7 Quranda interview.

8 Quranda interview.

9 Michael Hansen, executive director of Greater-Birmingham Alliance to Stop Pollution, interviewed by John M. Dunn, December 3, 2021.

10 Hansen interview.

11 Ramos, "Texas Energy Company."

12 Hanion, "Fusion Commercialization?"

13 Mumgaard, "CFS Will Build."
14 "Aura Aero."
15 Amscot, "Unlocking Arkansas' Hidden Treasure."
16 Lisa Hall, media & community relations advisor at Babcock Ranch, Florida, interviewed by John M. Dunn, April 9, 2024.
17 Hall interview.
18 Hall interview.

Twelve **Taking It to the Judge**

1 Lazarus, "Environmental Racism!"
2 Lazarus, "Environmental Racism!"
3 Chandra Taylor, senior attorney and leader of Southern Environmental Justice Initiative, Southern Environmental Law Center, Chapel Hill Office, interviewed by John M. Dunn, November 3, 2021.
4 Taylor interview.
5 Taylor interview.
6 Taylor interview.
7 "Climate Litigation Increasingly Reaching the Highest Courts around the world, Report Finds," Press release, Grantham Research Institute on Climate Change and the Environment at the London School of Economics and Political Science https://www.lse.ac.uk/granthaminstitute/news/climate-litigation-more-often-reaching-the-highest-courts-around-the-world-report-finds.
8 Volokh, "Foreign Law."
9 Tigre and Barry, Climate Change in the Courts.
10 Debre, "Youth Suing States."
11 Reynolds v. Florida Public Service Commission, 2024-019966-CA-01 / Fla. Cir. Ct. https://climatecasechart.com/wp-content/uploads/case-documents/2024/20241212_docket-2024-019966-CA-01_complaint.pdf.
12 Hersher, "Lawsuit."
13 City of Charleston v. Brabham Oil Co., 2020-CP-10-03975. https://www.charleston-sc.gov/ArchiveCenter/ViewFile/Item/716.
14 Traub, "Christopher Stone."
15 Ben Price, national organizing director of the Community Environmental Legal Defense Fund (CELDF), interviewed by John Dunn, December 3, 2021.
16 Price interview.

17 Price interview.
18 "Our History."

Thirteen The Point of No Return?

1 Godfrey, "'I Am Georgia.'"
2 Godfrey, "'I Am Georgia.'"
3 Moser, "Fearless Rise."
4 Sasso and Tanzi, "$100 Bill."
5 Sasso and Tanzi, "$100 Bill."
6 Chen, "Snapshot of Asian-Americans."
7 "Southeastern United States."
8 Renkl, "Lies, Damn Lies."
9 Alex Easdale, executive director of the Southeast Climate and Energy Network, interviewed by John Dunn, November 10, 2022.
10 Roberts, "Florida is Turning Its Back."
11 Roberts, "Florida is Turning Its Back."
12 Ripple et al., "World Scientists' Warning."
13 "Secretary-General's Press Conference on Climate by Secretary-General Antonio Guterres at United Nations Headquarters." Press release. United Nations, July 27, 2023. https://www.un.org/sg/en/content/sg/press-events/2023-07-27/secretary-generals-press-conference-climate.
14 "Press Conference by Secretary-General Antonio Guterres."
15 Taheri, "Millions of Americans."
16 Elliott, "New Book."
17 Laurence Tubiana (@LaurenceTubiana), X, November 6, 2024, https://x.com/LaurenceTubiana/status/1854115697267236999
18 Stefan Lehne, Senior Fellow at Carnegie Europe, email to John M. Dunn, December 11, 2022.
19 Lieven, *Climate Change*, xiv.
20 Lieven, *Climate Change*, xvi.
21 Lieven, *Climate Change*, 76.
22 Chiu, "He Got an Entire Country."
23 King, *Where Do We Go from Here?*
24 "Climate Change a Threat."

SOURCES

Adler, Ben, and Rebecca Leber. "Donald Trump Once Backed Urgent Climate Action. Wait, what?" *Vox*, June 9, 2016. https://www.vox.com/2016/6/9/11895578/donald-trump-climate-change.

Adler, Jonathan H. "The Republican Reversal." Cato Institute, Summer 2020. https://www.cato.org/regulation/summer-2020/Republican-reversal.

"Affinity Group Letter to David Yarnold and Executive Team, Audubon Society. People of Color Affinity Group, Early Career Professionals Affinity Group, LGBTQIA+ Affinity Group, Women's Employee Resource Group, Equity Diversity and Inclusion Working Group." https://www.politico.com/f/?id=00000175-ba7b-d1da-a775-ba7ff9a20000.

"Afton, NC (Warren County)." Real People—Real Stories. Exchange Project. Department of Health Behavior and Health Education, University of North Carolina at Chapel Hill, September 2006. https://exchangeproject.unc.edu/wp-content/uploads/sites/18406/2018/08/Real-People-Afton-long-story-07-0426.pdf.

Alston, Dana. "The Summit: Transforming a Movement." *RP&E Journal, Race, Poverty, and the Environment* 17, no. 1, Spring 2010. https://www.reimaginerpe.org/20years/alston.

"Alternative Summit Opens with Call for Climate Justice." CorpWatch, November 19, 2000. https://www.corpwatch.org/article/alternative-summit-opens-call-climate-justice.

Amscot, Sam. "Unlocking Arkansas' Hidden Treasure: USGS Uses Machine Learning to Show Large Lithium Potential in the Smackover Formation." U.S. Geological Survey, October 21, 2024, https://www.usgs.gov/news/national-news-release/unlocking-arkansas-hidden-treasure-usgs-uses-machine-learning-show-large.

Aning, Agya. "The Biden Administration's Embrace of Environmental Justice Has Made Wary Activists Willing to Believe." *Inside Climate News*, August 1, 2021, https://insideclimatenews.org/news/01082021/biden-environmental-justice.

Applebome, Peter. "The 1992 Campaign: The South; New Southern Strategy: Democrats Want It Back." *New York Times,* July 11, 1992, https://www.nytimes.com/1992/07/11/us/the-1992-campaign-the-south-new-southern-strategy-democrats-want-it-back.html.

Armey, Dick, and Matt Kibbe. *Give Us Liberty: A Tea Party Manifesto.* Harper Collins, 2010, pp. 90–91.

Asian Pacific Environmental Network: Who Is APEN? https://apen4ej.org/who-is-apen/.

"Aura Aero Inc. Selects Volusia County for 500,000 Square Foot Manufacturing and Assembly Plant in Florida." Embry-Riddle Aeronautical University News, October 17, 2024. https://news.erau.edu/headlines/aura-aero-inc-selects-volusia-county-for-manufacturing-and-assembly-plant-in-florida.

Ayres, Whit. "Quotation of the Day: How G.O.P. Shifted on Climate Science." *New York Times,* June 3, 2017. https://www.nytimes.com/2017/06/03/todayspaper/quotation-of-the-day-how-gop-shifted-on-climate-science.html.

Bakst, Daren. "Inflation Reduction Act Shows Far Left's Extremism and Elitism." Heritage Foundation, August 16, 2022. https://www.heritage.org/budget-and-spending/commentary/inflation-reduction-act-shows-far-lefts-extremism-and-elitism.

Bates, Sofie. "The Climate Events of 2020 Show How Excess Heat Is Expressed on Earth." NASA's Goddard Space Flight Center, January 14, 2021. https://www.nasa.gov/earth-and-climate/the-climate-events-of-2020-show-how-excess-heat-is-expressed-on-earth/.

Beals, Rachel Konig. "Experts Say UN's Next Climate Change Warning Will Be 'A Nightmare Painted in the Dry Language of Science.'" *Market Watch,* February 14, 2022. https://www.marketwatch.com/story/experts-say-uns-next-climate-change-warning-will-be-a-nightmare-painted-in-the-dry-language-of-science-11644854819.

Bean v. Southwestern Waste Management Corp., 482 F. Supp. 673—Dist. Court, SD Texas, 1979. https://scholar.google.com/scholar_case?case=3451324768386272698&q=,+Bean+V.+Southwestern+Waste+Management,&hl=en&as_sdt=40006&as_vis=1.

Benegal, Salil D. "The Spillover of Race and Racial Attitudes into Public Opinion about Climate Change." *Environmental Politics* 27, no. 4 (2018): 733–756. https://doi.org/10.1080/09644016.2018.1457287.

Berlin, Rubin N., and Gabrielle Wong-Parodi. "As California Burns: The Psychology of Wildfire- and Wildfire Smoke–Related Migration Intentions." *Population and Environment* 44 (August 24, 2022): 15–45. https://doi.org/10.1007/s11111-022-00409-w.

"Biden Administration Announces First-Ever Funding Program Dedicated to Reconnecting American Communities." Press release. U.S. Department of Transportation, June 30, 2022. https://www.transportation.gov/briefing-room/biden-administration-announces-first-ever-funding-program-dedicated-reconnecting.

"Billions of People Still Breathe Unhealthy Air: New WHO Data." World Health Organization, April 4, 2022. https://www.who.int/news/item/04-04-2022-billions-of-people-still-breathe-unhealthy-air-new-who-data.

"Biodiversity Survey Reveals That More Species Are Threatened with Extinction Than Previously Thought." *ScienceDaily*, July 18, 2022. https://www.sciencedaily.com/releases/2022/07/220718142320.htm.

Blakemore, Erin. "The Thorny History of Reparations in the United States." *History Channel*. September 14, 2023. https://www.history.com/news/reparations-slavery-native-americans-japanese-internment.

Bleby, Alice. "How 'Rights of Nature' Is Recasting the Relationship Between Law and the Earth." *Society of Environmental Journalists*, June 29, 2022. https://www.sej.org/publications/features/how-rights-nature-recasting-relationship-between-law-and-earth.

Block, Samuel. "Mining Energy-Transition Metals: National Aims, Local Conflicts." *MSCI*, June 3, 2021. https://www.msci.com/www/blog-posts/mining-energy-transition-metals/02531033947.

Bonnie, Robert, Emily Pechar Diamond, and Elizabeth Rowe. "Understanding Rural Attitudes Toward the Environment and Conservation in America," NI R 20-03, Duke University. https://nicholasinstitute.duke.edu/sites/default/files/publications/understanding-rural-attitudes-toward-environment-conservation-america.pdf.

Braimah, A. "Houston (TSU) Riot 1967." BlackPast.org, University of Washington, December 4, 2017. https://www.Blackpast.org/african-american-history/houston-tsu-riot-1967/.

Britton-Purdy, Jedediah. "Environmentalism Was Once a Social-Justice Movement: It Can Be Again." *Atlantic*, December 7, 2016. https://www.theatlantic.com/science/archive/2016/12/how-the-environmental-movement-can-recover-its-soul/509831/.

Brodkin, Karen. *Environmental Activism in South Los Angeles*. Rutgers University Press, 2009.

Bruggers, James. "Southern State Energy Officials Celebrate Fossil Fuels as World

Raises Climate Alarm." *Inside Climate News,* September 26, 2019. https://insideclimatenews.org/news/26092019/fossil-fuels-climate-change-southern-states-energy-board-koch-kentucky/.

Brune, Michael. "Pulling Down Our Monuments." Sierra Club, July 22, 2020. https://www.sierraclub.org/michael-brune/2020/07/john-muir-early-history-sierra-club.

Bryant, Bunyan, and Paul Mohai. "The Michigan Conference: A Turning Point." *EPA Journal* 18, no. 1, March/April 1992.

Buford, Talia. "Has the Moment for Environmental Justice Been Lost?" *ProPublica,* July 24, 2017. https://www.propublica.org/article/has-the-moment-for-environmental-justice-been-lost.

Bullard, Robert D. "Environmental Justice for All." Nature Transformed, the Environment in American History. *TeacherServe.* National Humanities Center. https://nationalhumanitiescenter.org/tserve/nattrans/ntuseland/essays/envjust.htm.

Bullard, Robert D. "Environmental Justice in the 21st Century: Race Still Matters." *Phylon* 49, no. 3/4. Clark Atlanta University. Autumn-Winter, 2001. https://www.jstor.org/stable/3132626.

Bullard, Robert D., Paul Mohai, Robin Saha, and Beverly Wright. "Toxic Wastes and Race at Twenty: 1987–2007." A Report Prepared for the United Church of Christ Justice & Witness Ministries. United Church of Christ. https://www.ucc.org/wp-content/uploads/2021/03/toxic-wastes-and-race-at-twenty-1987-2007_Part1.pdf.

Bullock, Charles S. III, Susan A. MacManus, Jeremy D. Mayer, and Mark J. Rozell. *The South and the Transformation of U.S. Politics.* Oxford University Press, 2019.

Burbank, Luke. "Bush Views Shift on Climate Change." National Public Radio, February 1, 2007. https://www.npr.org/templates/story/story.php?storyId=7115660.

Burwell, Dollie, and Luke W. Cole. "Environmental Justice Comes Full Circle: Warren County Before and After." *Golden Gate University Environmental Law Journal* 13, no. 1 (2007). http://digitalcommons.law.ggu.edu/gguelj/vol1/iss1/4.

"The Bush Administration's Misuse of Science." Remarks to the National Academy of Sciences. Union of Concerned Scientists, April 23, 1990. http://www.webexhibits.org/bush/1.html.

Bush, George H. W. "The President's News Conference in Rio de Janeiro." *American Presidency Project,* University of California Santa Barbara, June 13, 1992. https://www.presidency.ucsb.edu/documents/the-presidents-news-conference-rio-de-janeiro.

Bustamante, Roxanne. "I-45 Expansion Project Remains Stalled Due to Federal

Investigation." ABC 13 Houston News, April 2, 2022. https://abc13.com/i-45-expansion-project-remains-stalled-the-harris-county-attorneys-office-txdot/11701692/.

Byrnes, Joe. "West Ocala Community Center Opens Where a Charcoal Plant Once Spewed Soot over African American Homes." *Central Florida Public Media,* January 12, 2023. https://www.cfpublic.org/environment/2023-01-10/west-ocala-community-center-opens.

Carnel, Zachary, and Isabella Douglas. "Remembering Rosewood: Descendants Mark Racial Violence That Razed Florida Town 100 Years Ago." WUFT, a PBS member television station in Gainesville, Florida. January 3, 2023. https://www.wuft.org/human-interest/2023-01-03/remembering-rosewood-descendants-mark-racial-violence-that-razed-florida-town-100-years-ago.

Carroll, Sean B. "The Denialist PlayBook." *Scientific American,* November 8, 2020. https://www.scientificamerican.com/article/the-denialist-playbook.

Chang, Jeff, and Lucia Hwang. "It's A Survival Issue: The Environmental Justice Movement Faces the New Century." *ColorLines* 3, no. 2, Summer 2000.

"Charleston, SC Is First Southern City to Sue Big Oil for Climate Change Damages." Center for Climate Integrity, September 9, 2020. https://climateintegrity.org/news/charleston-sc-is-first-southern-city-to-sue-big-oil-for-climate-change-damages.

Chen, Angela. "A Snapshot of How Asian-Americans Are Changing the South." *Huffington Post,* March 28, 2018. https://www.huffpost.com/entry/asian-american-growth-south_n_5a948133e4b0699553cb5834.

Cheun, Phil. "GOP and Democratic Platforms Highlight Stark Differences on Energy and Climate." *Inside Climate News,* July 26, 2016. https://insideclimatenews.org/news/26072016/democrat-Republican-party-platforms-energy-climate-change-hillary-clinton-donald-trump.

Cheung, Helier. "What Does Trump Actually Believe on Climate Change?" *BBC News,* January 23, 2020. https://www.bbc.com/news/world-us-canada-51213003.

"A Chilling Story About Climate Change." Editorial, *Concord Monitor,* August 26, 2018. https://www.concordmonitor.com/2018/08/26/sununu-and-climate-change-19676534/.

Chiu, Allyson. "He Got an Entire Country Running on Clean Energy. Can He Do It Again?" *Washington Post,* September 21, 2025. https://www.washingtonpost.com/climate-solutions/2025/09/19/uruguay-renewable-energy-climate-breakthrough.

Cho, Renée. "Why Climate Change Is an Environmental Justice Issue." State of the Planet. Columbia Climate School, September 22, 2020. https://news.climate.columbia.edu/2020/09/22/climate-change-environmental-justice.

Choi, Christina. "Why the Environmental Movement Should Stop Ignoring Asian Americans." National Resources Defense Council, May 3, 2021. https://www.nrdc.org/stories/why-environmental-movement-should-stop-ignoring-asian-americans.

City of Charleston v. Brabham Oil Co. 2020-CP-10-03975. https://climateintegrity.org/uploads/media/2020-09-09-Charleston-Climate-Complaint.pdf.

City of Houston v. George. 479 S.W.2d 257 (1972). https://law.justia.com/cases/texas/supreme-court/1972/b-2726-0.html.

"Climate Change a Threat to Human Wellbeing and Health of the Planet." *IPCC Sixth Assessment Report: Impact, Adaptation and Vulnerability,* February 28, 2022. https://www.ipcc.ch/report/ar6/wg2/resources/press/press-release.

"Climate Change in Texas." Public Citizen, 2023. https://www.citizen.org/climate-change-in-texas/.

"Climate Change Laws of the World." Grantham Research Institute on Climate Change and the Environment, London School of Economics and Political Science. https://www.lse.ac.uk/granthaminstitute/climate-change-laws-of-the-world-database.

"Climate Change Research Distorted and Suppressed." Reports & Multimedia/Case Study, Union of Concerned Scientists, June 30, 2005. https://www.ucsusa.org/resources/climate-change-research-distorted-and-suppressed.

"Climate Justice." *Stanford Encyclopedia of Philosophy.* Stanford University, June 4, 2020. https://plato.stanford.edu/entries/justice-climate/.

"Climate Justice." Sustainable Development Goals, United Nations, May 31, 2019. https://www.un.org/sustainabledevelopment/blog/2019/05/climate-justice/.

"Climate Justice in Rural Southeastern United States: A Review of Climate Change Impacts and Effects on Human Health." National Library of Medicine, National Center for Biotechnology Information, February 13, 2016. https://www.ncbi.nlm.nih.gov/pmc/articles/PMC4772209.

"Climate Litigation Increasingly Reaching the Highest Courts around the world, Report Finds." Press release, Grantham Research Institute on Climate Change and the Environment, London School of Economics and Political Science. https://www.lse.ac.uk/granthaminstitute/news/climate-litigation-more-often-reaching-the-highest-courts-around-the-world-report-finds.

"Climate Mayors Letter to U.S. House of Representatives, Speaker Nancy Pelosi, U.S. Senate Majority Leader Mitch McConnell, U.S. House Minority Leader Kevin McCarthy, U.S. Senate Minority Leader Chuck Schumer." Climate Mayors, July 21, 2020. https://static1.squarespace.com/static/6559a0eea511ed3e8294b547/t

/666370f3f78a5d35ceecbeae/1717793023011/Climate-Mayors_Congressional-Letter_R5-compressed.pdf.
"Climate Mayors National Dialogue on Green and Equitable Recovery Southeast Region." Climate Mayors, July 23, 2020. https://www.youtube.com/watch?v=TKDI1GcGxTE.
Collomb, Jean-Daniel. "The Ideology of Climate Change Denial in the United States." *European Journal of American Studies*, Spring 2014. https://journals.openedition.org/ejas/10305.
Colman, Zack. "Environmental Groups' Greatest Obstacle May Not Be Republican Opposition." *Politico*, February 5, 2021. https://www.politico.com/news/magazine/2021/02/05/environmental-movement-racial-reckoning-green-diversity-465501.
"Come Hell or High Water, *The Battle for Turkey Creek*." A documentary by Leah Mahan, Bullfrog Films. Zamier Productions, Oley, PA, 2013.
Cost, Ben. "Nickelodeon's 'Environmental Racism' Segment Sparks Backlash from Conservative Viewers." *New York Post*, April 26, 2021. https://nypost.com/2021/04/26/conservatives-slime-nickelodeons-environmental-racism-segment/,
Covar, Jak. Letter from Jak Covar, Texas Commission on Environmental Quality, to Gerald Filbin, U.S. Environmental Protection Agency, January 3, 2014. https://www.tceq.texas.gov/assets/public/agency/nc/multimedia/TCEQ-Comments-on-EPA-HQ-2013-0568.pdf.
Cox, Jeremy. "Progress on Environmental Justice Draws Criticism in Maryland." *Bay Journal*, October 12, 2021. https://www.bayjournal.com/news/people/progress-on-environmental-justice-draws-criticism-in-maryland/article_92a86904-2148-11ec-913b-a30436a05f91.html.
Cramton, Peter. "Fostering Resiliency with Good Market Design: Lessons from Texas." Cramton Papers at the University of Maryland, January 2022. http://www.cramton.umd.edu/papers2020-2024/cramton-lessons-from-the-2021-texas-electricity-crisis.pdf.
Cusick, Daniel, and *E&E News*. "This Part of the U.S. Will Suffer Most from Climate Change." *Climate Wire, Scientific American*, April 5, 2023. https://www.scientificamerican.com/article/this-part-of-the-u-s-will-suffer-most-from-climate-change/
Daniel, Pete. *Standing at the Crossroads, Southern Life in the Twentieth Century.* Johns Hopkins University Press, 1996.
Davenport, Coral, and Eric Lipton. "How G.O.P. Leaders Came to View Climate Change as Fake Science." *New York Times*, June 3, 2017. https://www.nytimes.com/2017/06/03/us/politics/republican-leaders-climate-change.html.

Davidson, Curt. "Emelle, Alabama: Home of the Nation's Largest Hazardous Waste Landfill." *History of Environmental Justice: Education and Research at School for Environment and Sustainability.* University of Michigan, 2023. https://web.archive.org/web/20170320023532/http://umich.edu/~snre492/Jones/emelle.htm.

Debre, Elena. "Youth Suing States over Climate Change Will Have Their Day in Court, and Public Opinion Is on Their Side." News & Events, Climate Change Communication. Yale University, April 20, 2022. https://climatecommunication.yale.edu/news-events/youth-climate-lawsuits-in-montana-oregon-and-virginia-demonstrate-ypccc-findings.

Denchak, Melissa. "Paris Climate Agreement: Everything You Need to Know." National Resources Defense Council, February 19, 2021. https://www.nrdc.org/stories/paris-climate-agreement-everything-you-need-know.

"Denial, Disinformation, and Doublespeak: Big Oil's Evolving Efforts to Avoid Accountability for Climate Change." Joint Staff Report, Democratic Staff of the House Committee on Oversight and Accountability, April 2024. https://www.budget.senate.gov/imo/media/doc/denial__disinformation_and_doublespeak_big_oils_evolving_efforts_to_avoid_accountability_for_climate_change.pdf.

Desai, Mihir, Suzanne Antonio, Leanne Fan, and the HBS Technology Group. "The Tulsa Massacre and the Call for Reparations." Harvard Business School. https://store.hbr.org/product/the-tulsa-massacre-and-the-call-for-reparations-multimedia-case/134HBD.

Dickey, Colin. *Under the Eye of Power: How Fear of Secret Societies Shapes American Democracy.* Viking, 2023.

Dillen, Abigail. "What We Saw at COP2: Lackluster Outcomes but Hopeful Connections." Earthjustice, November 21, 2022. https://earthjustice.org/experts/abigail-dillen/what-we-saw-at-cop27-lackluster-outcomes-but-hopeful-connections.

Dillon, Jeremy. "House Republicans Prepare Big Energy Package for 2023." *E&E-News By Politico,* November 18, 2022. https://www.eenews.net/articles/house-Republicans-prepare-big-energy-package-for-2023/.

"Disparities in the Impact of Air Pollution." American Lung Association, November 2, 2023. https://www.lung.org/clean-air/outdoors/who-is-at-risk/disparities.

Douglas, Erin. "EPA Will Center Climate Change Response in Texas on Sea Level Rise, Floods, Drought and Severe Storms." *Texas Tribune,* October 6, 2022. https://www.texastribune.org/2022/10/06/epa-plan-texas-climate-change/.

Douglas, Erin. "Texas' Plan to Provide Water for a Growing Population Virtually Ignores Climate Change." *Texas Tribune,* October 31, 2022. https://www.texastribune.org/2022/10/31/texas-water-plan-reservoirs-climate-change/.

Drennen, Ari, and Sally Hardin. "Climate Deniers in the 117th Congress." Center for American Progress, March 30, 2021. https://www.americanprogress.org/article/climate-deniers-117th-congress/.

D'Souza, Deborah. "Understanding the Green New Deal & What's in the Climate Proposal: A Call to End Fossil Fuels and Build Green Jobs." *Investopedia,* May 28, 2022. https://www.investopedia.com/the-green-new-deal-explained-4588463.

Dumain, Emma. "Republicans Deploy Old and New Tactics Against EJ Bill." *E&E-News By Politico,* February 16, 2022. https://www.eenews.net/articles/Republicans-deploy-old-and-new-tactics-against-ej-bill/.

Dunlap, Riley E., and Aaron M. McCright. "A Widening Gap: Republican and Democratic Views on Climate Change." *Environment Science and Policy for Sustainable Development* 50, no. 5 (2008): 26–35. https://www.researchgate.net/publication/241737833_A_Widening_Gap_Republican_and_Democratic_Views_on_Climate_Change

Durlin, Marty. "The Group of 10 Respond." *High Country News,* February 1, 2010.

Durlin, Marty. "The Shot Heard Round the West." *High Country News,* February 1, 2010. https://www.hcn.org/issues/42.2/the-shot-heard-round-the-west.

Edge, Thomas. "Southern Strategy 2.0: Conservatives, White Voters, and the Election of Barack Obama." *Journal of Black Studies* 40, no. 3 (2010). Southern Strategy 2.0: JSTOR.

Elliott, Debbie. "A New Book Argues the U.S. South Shapes the Nation's Political and Cultural Landscape." National Public Radio, March 26, 2022. https://www.npr.org/2022/03/26/1088991099/a-new-book-argues-the-u-s-south-shapes-the-nations-political-and-cultural-landsc.

Ellis, Ralph, and Hollie Silverman, "'Tidal Wave' of Floodwaters Overtook Tennesseans in Minutes and Killed 21, Including 7-month-old Twins." *CNN,* August 23, 2021. https://www.cnn.com/2021/08/23/us/tennessee-fatal-flooding-monday/index.html.

"Environmental Justice." United States Environmental Protection Agency. https://www.epa.gov.

"The EPA Starts a Civil Rights Probe into Mississippi's Water Woes." Associated Press: NPR, October 20, 2022. https://www.npr.org/2022/10/20/1130393118/the-epa-starts-a-civil-rights-probe-into-mississippis-water-woes.

"Exposed: Social Vulnerability and Climate Change in the US Southeast." Oxfam America, 2. https://s3.amazonaws.com/oxfam-us/www/static/oa3/files/Exposed-Social-Vulnerability-and-Climate-Change-in-the-US-Southeast.pdf.

"Extinction Rebellion Rebels Against Extinction in Atlanta, Worldwide." *Atlanta Pro-*

gressive, May 17, 2019. https://atlantaprogressivenews.com/2019/05/17/extinction-rebellion-rebels-against-extinction-in-atlanta-worldwide/.

"Fact Sheet: A Year Advancing Environmental Justice." Press release. The White House, January 26, 2022. Gerhard Peters and John T. Woolley, The American Presidency Project, https://www.presidency.ucsb.edu/node/354261.

"Fact Sheet: Inflation Reduction Act Advanced Environmental Justice." Press release. The White House, August 17, 2022. https://bidenwhitehouse.archives.gov/briefing-room/statements-releases/2022/08/17/fact-sheet-inflation-reduction-act-advances-environmental-justice.

"Fact Sheet: President Biden Takes Executive Actions to Tackle the Climate Crisis at Home and Abroad, Create Jobs, and Restore Scientific Integrity Across Federal Government." Press release. The White House. January 27, 2021. Gerhard Peters and John T. Woolley, The American Presidency Project, https://www.presidency.ucsb.edu/node/347887.

Falcon, Russell, and Nexstar Media Wire. "The 10 Most Polluted States in the US." *The Hill,* September 10, 2022. https://thehill.com/policy/equilibrium-sustainability/3624274-the-10-most-polluted-states-in-the-us.

Fears, Darryl, and Brady Dennis. "This Is Environmental Racism." *Washington Post,* April 6, 2021. https://www.washingtonpost.com/climate-environment/interactive/2021/environmental-justice-race/?itid=sr_0_4329c3c8-5d8b-4fe8-9661-e19bc721fa76.

Ferman, Mitchell. "Texas Bans Local, State Government Entities from Doing Business with Firms that 'Boycott' Fossil Fuels." *Texas Tribune,* August 24, 2022. https://www.texastribune.org/2022/08/24/texas-boycott-companies-fossil-fuels/

Ferris, Deeohn. "Environmental Justice: Moving Equity from Margins to Mainstream." *Nonprofit Quarterly,* August 15, 2019. https://nonprofitquarterly.org/environmental-justice-moving-equity-from-margins-to-mainstream.

Ferris, Deeohn. "Say This, Not That: Why Big Green Nonprofits Must Embrace Racial Equity." *Nonprofit Quarterly,* August 10, 2020. https://nonprofitquarterly.org/say-this-not-that-why-big-green-nonprofits-must-embrace-racial-equity.

"55 Arrested in Protests at a Toxic Dump in Carolina." Special to the *New York Times,* September 16, 1982. https://www.nytimes.com/1982/09/16/us/55-arrested-in-protest-at-a-toxic-dump-in-carolina.html.

"First in the U.S.: 'Rights of Nature' State Constitutional Amendment Filed in Florida to Protect Waterways." Center for Democratic and Environmental Rights, June 2, 2021. https://www.centerforenvironmentalrights.org/news/first-in-the-us-rights-of-nature-state-constitutional-amendment-filed-in-florida-to-protect-waterways.

Florida Conservation Voters, "About Chispa." https://chispalcv.org/about-us/
"For Wetland Plants, Sea-Level Rise Stamps Out Benefits of Higher CO2." Smithsonian Environmental Research Center, May 18, 2022. https://www.si.edu/newsdesk/releases/wetland-plants-sea-level-rise-stamps-out-benefits-higher-co2.
Foreman, Christopher H., Jr. "The Clash of Purposes: Environmental Justice and Risk Assessment." Brookings Institution, March 20, 1998. https://www.brookings.edu/articles/the-clash-of-purposes-environmental-justice-and-risk-assessment/.
Foreman, Christopher H., Jr. "Facing Up to Racial Disparity: Daunting Realities Hinder the Drive for Equality." Brookings Institution, March 1, 2000, https://www.brookings.edu/articles/facing-up-to-racial-disparity-daunting-realities-hinder-the-drive-for-equality.
Foreman, Christopher H., Jr. "Remarks on Environmental Justice." Testimony to the United States Commission on Civil Rights. Brookings Institution, January 11, 2002. https://www.brookings.edu/testimonies/remarks-on-environmental-justice/.
Foreman, Christopher H., Jr. "A Winning Hand? The Uncertain Future of Environmental Justice." Brookings Institution, March 1, 1996. https://www.brookings.edu/articles/a-winning-hand-the-uncertain-future-of-environmental-justice.
Friedman, David. "The 'Environmental Racism' Hoax." *American Enterprise* 9, no. 6 (1998): 75+. https://link.gale.com/apps/doc/A21249963/AONE.
Friedman, Thomas L. "Opinion—A Warning from the Garden." *New York Times*, January 19, 2007. https://www.nytimes.com/2007/01/19/opinion/19friedman.html.
Fritze, John, David Jackson, and Michael Collins. "Critics Slam Trump 'Suburban Housewife' Tweet as Racist, Sexist 'Dog Scream' Play for White Voters," *USA Today*, August 12, 2020. https://www.usatoday.com/story/news/politics/elections/2020/08/12/trump-critics-see-tweet-the-suburban-housewife-sexist-racist/3348444001/.
"Frontline Values for a Green New Deal." Gulf South for a Green New Deal Policy Platform, 2019. https://www.gulfsouth4gnd.org/_files/ugd/a491a1_637aea05b7814ad5b917daca77777118.pdf.
Funk, Cary, and Meg Hefferon. "U.S. Public Views on Climate and Energy." Pew Research Center, November 25, 2019. https://www.pewresearch.org/science/2019/11/25/u-s-public-views-on-climate-and-energy.
[GAO] US Government Accounting Office. Siting of Hazardous Waste Landfills and Their Correlation with Racial and Economic Status of Surrounding Communities (GAO/RECD-83-168), Washington, DC: GAO, 1983. https://www.gao.gov/assets/rced-83-168.pdf.
Garamome, Jim. "DOD Begins Implementing Naming Commission Recommendations." *DOD News*. U.S. Department of Defense, January 5, 2023. https://

www.defense.gov/News/News-Stories/Article/article/3260434/dod-begins-implementing-naming-commission-recommendations.

Gardiner, Beth. "Unequal Impact: The Deep Links Between Racism and Climate Change." Yale University. *YaleEnvironment 360,* June 9, 2020.

Gelles, David, and Hirdco Tabuchi. "How an Organized Republican Effort Punished Companies for Climate Action." *New York Times,* May 27, 2022. https://www.nytimes.com/2022/05/27/climate/republicans-Blackrock-climate.html.

Gil, Bruce. "U.S. Cities and States Are Suing Big Oil over Climate Change. Here's What the Claims Say and Where They Stand." *Frontline.* Public Broadcasting System, August 1, 2022. https://www.pbs.org/wgbh/frontline/article/us-cities-states-sue-big-oil-climate-change-lawsuits.

Glickman, Lawrence. "How White Backlash Controls American Progress." *Atlantic,* May 22, 2020. https://www.theatlantic.com/ideas/archive/2020/05/white-backlash-nothing-new/611914/.

Godfrey, Gavin. "I Am Georgia: Read Raphael Warnock's Victory Speech," Warnock Acceptance Speech. Capital B Atlanta, December 7, 2022. https://atlanta.capitalbnews.org/warnock-victory-speech/.

Goldenberg, Suzanne. "The Worst of Times: Bush's Environmental Legacy Examined." *Guardian,* January 16, 2009. https://www.theguardian.com/politics/2009/jan/16/greenpolitics-georgebush.

"Governor Cooper Signs Executive Order Detailing Next Steps on Path to a Clean Energy and Equitable Economy for All North Carolinians." Press Release. Governor's Office, January 7, 2022. https://governor.nc.gov/news/press-releases/2022/01/07/governor-cooper-signs-executive-order-detailing-next-steps-path-clean-energy-and-equitable-economy.

Grady, Henry. "The New South." Speech given to the New England Society in New York City. December 21, 1886. *The American Yawp Reader,* ed. Joseph L. Locke. Stanford University Press. https://www.americanyawp.com/reader/18-industrial-america/henry-grady-on-the-new-south-1886.

Grant, Richard. *The Deepest South of All: True Stories from Natchez, Mississippi.* Simon & Schuster, 2020.

"Green New Deal." Sierra Club. https://www.sierraclub.org/green-new-deal.

Greensboro Truth & Reconciliation Commission Final Report, May 25, 2006. https://hmcwordpress.humanities.mcmaster.ca/Truthcommissions/wp-content/uploads/2019/01/UnitedStates.Greensboro.TRC_.Report-FULL.pdf.

Griffith, Keith. "Economist Proposes Slavery Reparations of up to $12 Trillion—

Equivalent to $800,000 per Black Household or More Than Half of the American GDP," *Daily Mail,* August 13, 2020. https://www.dailymail.co.uk/news/article-8624985/Economist-proposes-slavery-reparations-12-TRILLION.html

Gross, Samantha J., and Alex Harris. "Key Biscayne Needs to Adapt to Climate Change. Will Voters OK $100M to Fight It?" Miami Herald, September 03, 2020. https://www.miamiherald.com/news/local/community/miami-dade/key-biscayne/article245378270.html.

Haney López, Ian. "Ian Haney López on the Dog Whistle Politics of Race (Part I)." Moyers & Company, February 28, 2014. https://billmoyers.com/episode/ian-haney-lopez-on-the-dog-whistle-politics-of-race.

Hanion, Elizabeth. "Is Fusion Commercialization in Sight? Not Yet, Says John Holdren." Harvard Kennedy School Belfer Center's Science, Technology, and Public Policy Program. April 3, 2004. https://www.belfercenter.org/publication/fusion-commercialization-sight-not-yet-says-john-holdren.

Hannah-Jones, Nikole, Caitlin Roper, Ilena Silverman, and Jake Silverstein, eds. *The 1619 Project.* One Word, Penguin Random House, 2021.

Harris, Richard. "'Uncertain' Science: Judith Curry's Take on Climate Change." National Public Radio, August 22, 2013. https://www.npr.org/2013/08/22/213894792/uncertain-science-judith-currys-take-on-climate-change.

Hassan, Adeel, and Jack Healy. "America Has Tried Reparations Before. Here Is How It Went." *New York Times,* June 19, 2019. https://www.nytimes.com/2019/06/19/us/reparations-slavery.html

Hauer, Matthew. "Migration Induced by Sea Level Rise Could Reshape the U.S. Population Landscape." *Nature Climate Change 7* (2017): 321–325. https://doi.org/10.1038/nclimate3271.

Hayhoe, Katherin, with Dave Strauss. Nature Conservancy tele-townhall presentation. September 2025.

Heer, Jeet. "How the Southern Strategy Made Donald Trump Possible." *New Republic,* February 18, 2016. https://newrepublic.com/article/130039/southern-strategy-made-donald-trump-possible.

Heersink, Boris, and Jeffrey A. Jenkins. "The Republican Party Is White and Southern. How Did That Happen?" *Washington Post,* February 7, 2020. https://www.washingtonpost.com/politics/2020/02/07/republican-party-is-white-southern-how-did-that-happen.

Helman, Christopher. "America's 20 Worst Corporate Air Polluters." *Forbes,* June 11, 2013. https://www.forbes.com/sites/christopherhelman/2013/06/10/americas-20-worst-corporate-air-polluters/.

Henderson, Todd, and Jeremy Kidd. "Doable Reparations for Native Americans." *Newsweek*, October 14, 2021. https://www.newsweek.com/doable-reparations-native-americans-opinion-1638566.

Heritage, John. "From the Editorial Page." *EPA Journal* 8, no. 1, March/April 1992.

Hersher, Rebecca. "Lawsuit Alleging Oil Companies Misled Public About Climate Change Moves Forward." National Public Radio, January 25, 2022. https://www.npr.org/2022/01/25/1075560742/oil-lawsuit-climate-change-baltimore.

Hersher, Rebecca. "U.S. Government to Pay $492 Million to 17 American Indian Tribes." The Two Way. National Public Radio, September 27, 2016. https://www.npr.org/sections/thetwo-way/2016/09/27/495627997/u-s-government-to-pay-492-million-to-17-american-indian-tribes.

Higgs, Nicolette. "How Green Theology Is Energizing the Black Community to Fight the Climate Crisis." *CNN*, November 22, 2019. https://www.cnn.com/2019/11/22/us/green-the-church-black-community-fights-climate-change/index.html.

"History of Environmental Justice and Research at SEAS." SEAS School for Environmental and Sustainability. University of Michigan. https://seas.umich.edu/academics/master-science/environmental-justice/history-environmental-justice.

Hofstadter, Richard. *Anti-intellectualism in American Life*. Alfred A, Knopf, 1964.

Holcomb, Sarah. "How 1,000 Black Churches Are Caring for the Earth: A Conversation with Kim Noble of Green the Church." *Christianity Today*, April 22, 2021.

Horning, Gloria G. "Social Network and Environmental Justice: A Case Study in Perry, Florida," 2005. https://repository.lib.fsu.edu/islandora/object/fsu%3A182023.

"Houston Climate Action Plan." Mayor Sylvester Turner. City of Houston. http://greenhoustontx.gov/climateactionplan/CAP-April2020.pdf.

"How Religious Is Your State?" Pew Research Center. February 29, 2016. https://www.pewresearch.org/short-reads/2016/02/29/how-religious-is-your-state/?state=alabama.

Howard, Pat. "Best and Worst States for Climate Change." Policygenius, October 5, 2022. https://www.policygenius.com/homeowners-insurance/best-and-worst-states-for-climate-change.

"The Impact of Climate Change on U.S. Subnational Economies." Moody's Analytics. https://www.moodys.com/web/en/us/insights/physical-transition-risk/the-impact-of-climate-change-on-us-subnational-economies.html.

Irfan, Ulmair. "Scientists Are Divided over Whether Climate Change Is Fueling Extreme Cold Events." *Vox*, February 18, 2021. https://www.vox.com/22287295/texas-uri-climate-change-cold-polar-vortex-arctic.

Jackson, John J. "The Southernization of American Politics." Remarks made at the Arkansas Political Science Association Convention, February 1986. https://journals.shareok.org/arp/article/view/848/801.

Jacques, Peter J., Riley E. Dunlap, and Mark Freeman. "The Organisation of Denial: Conservative Think Tanks and Environmental Scepticism." *Environmental Politics* 17, no. 3 (2008): 349–385. https://doi.org/10.1080/09644010802055576.

"James Butler and the Flood of 1927." *American Experience.* Public Broadcasting System. https://www.pbs.org/wgbh/americanexperience/features/neworleans-james-butler-and-flood-1927.

Jeffreys, Kent. "Environmental Racism: A Skeptic's View." *Journal of Civil Rights and Economic Development* 9, no. 2, Spring 1994, Article 27. https://scholarship.law.stjohns.edu/cgi/viewcontent.cgi?article=1484&context=jcred/.

Jones, Jeffrey, "Americans Say Government Should Address Slavery Effects," Gallup News. Gallup Center. June 16, 2022. https://news.gallup.com/poll/393689/americans-say-government-address-slavery-effects.aspx.

Jones, Rachel. "The Environmental Movement Is Very White. These Leaders Want to Change That." National Geographic News, July 29, 2020. https://www.nationalgeographic.com/history/article/environmental-movement-very-white-these-leaders-want-change-that.

Jones, Robert P. *The Hidden Roots of White Supremacy and the Path to a Shared American Future.* Simon & Schuster, 2023.

Kamins, Adams. "The Impact of Climate Change on U.S. Subnational Economies." Moody's Analytics, February 2023. https://www.moodys.com/web/en/us/insights/physical-transition-risk/the-impact-of-climate-change-on-us-subnational-economies.html.

Kaminski, Isabella. "How Scientists Are Helping Sue over Climate Change." *Lancet, Planetary Health,* May 2022. https://www.thelancet.com/journals/lanplh/article/PIIS2542-5196(22)00098-5/fulltext.

Karliner, Joshua. "Climate Justice Summit Provides Alternative Vision." CorpWatch, November 21, 2000. https://www.corpwatch.org/article/climate-justice-summit-provides-alternative-vision.

Karst, James. "The Mid-Barataria Sediment Diversion: Fact vs. Fiction." Delta Dispatches. RESTORE The Mississippi River Delta. September 20, 2021. https://mississippiriverdelta.org/the-mid-barataria-sediment-diversion-fact-vs-fiction.

Katz, Cheryl. "Birth of an Environmental Movement: Q&A with Pioneers." Climate Central, June 27, 2012. https://www.climatecentral.org/news/birth-of-the-movement-a-qa-with-two-environmental-justice-pioneers.

Kimmell, Ken, and Kathleen Rest. "An Open Letter to Supporters and Partners." Union of Concerned Scientists, June 5, 2020.

King, Martin Luther, Jr. *Where Do We Go from Here: Chaos or Community?* New York: Harper & Row, 1967.

Krawzak, Paul M. "Trump Administration Veteran Shops Budget Plan with Deep Cuts." *Roll Call,* December 7, 2022. https://rollcall.com/2022/12/07/trump-administration-veteran-shops-budget-plan-with-deep-cuts.

Kromm, Chris. "As the South Goes . . ." Facing South. Institute for Southern Studies, December 16, 2014. https://www.facingsouth.org/2014/12/as-the-south-goes.html.

Labalme, Jenny. "From the Archives: Dumping on Warren County." Facing South, September 30, 2022. https://www.facingsouth.org/2022/09/archives-dumping-warren-county.

Laporte, Amaury. "Remembering George H. W. Bush, the 'Environmental President.'" Environmental and Energy Study Institute, December 5, 2018. https://www.eesi.org/articles/view/remembering-george-h.w.-bush-the-environmental-president.

"Latinos Communities and Climate Change. Why We Care and What We Can Do." Environmental Defense Fund. https://www.edf.org/sites/default/files/content/latinos_and_climate_change_factsheet_0317_refresh.pdf.

Lavoie, Denise. "Virginia Judge Dismisses Youth Climate Change Lawsuit on Fossil Fuel Projects." Associated Press, Yahoo News, September 16, 2022. https://www.yahoo.com/news/virginia-judge-dismisses-youth-climate-205740466.html.

Lazarus, Richard J. "Environmental Racism! That's What It Is." *University of Illinois Law Review* 2000, pp. 255–274. Reprinted by Georgetown University Law Center, Georgetown Law Faculty Publications, January 2010. https://scholarship.law.georgetown.edu/facpub/160.

Lehne, Stefan. "What the COVID-19 Pandemic Tells Us About Climate Change and Diplomacy." Carnegie Europe. https://carnegieendowment.org/research/2021/10/what-the-covid-19-pandemic-tells-us-about-climate-change-and-diplomacy.

"The Lewis Powell Memo: A Corporate Blueprint to Dominate Democracy." Greenpeace, August 23, 1971. https://www.greenpeace.org/usa/democracy/the-lewis-powell-memo-a-corporate-blueprint-to-dominate-democracy.

Lieven, Anatol. *Climate Change and the Nation State: The Case for Nationalism in a Warming World.* New York: Oxford University Press, 2020.

Lind, Michael. *Made in Texas: George W. Bush and the Southern Takeover of American Politics.* New York: Basic Books, 2013.

Lindsey, Rebecca, Luannn Dahlman, and Jessica Blinden. "Climate Change: Global Temperature." NOAA National Centers for Environmental Information, January 18, 2023. https://www.climate.gov/news-features/understanding-climate/climate-change-global-temperature.

Love, Shayla. "Southerners Are Scared of the Climate Crisis, and Their Politicians Are Ignoring Them." *VICE News*, July 16, 2019. https://www.vice.com/en/article/southerners-are-scared-of-the-climate-crisis-and-their-politicians-are-ignoring-them.

Ludden, Jennifer. "Avoiding the Tap Water in Jackson, Miss. Has Been a Way of Life for Decades." National Public Radio, September 4, 2022. https://www.npr.org/2022/09/04/1120887065/jackson-mississippi-race-water-divide-politics.

MacLean, Nancy. *Democracy in Chains: The Deep History of the Radical Right's Stealth Plan for America.* New York: Viking, 2017.

Maloney, Thomas N. "African Americans in the Twentieth Century." Economic History Association. https://eh.net/encyclopedia/african-americans-in-the-twentieth-century.

"The Many Faces of Climate Justice: Exploring the Principles of Climate Justice." Mary Robinson Foundations Climate Justice. https://www.mrfcj.org/our-work/the-many-faces-of-climate-justice-exploring-the-principles-of-climate-justice/.

Marcetic, Branko. "The Democrats Are Climate Deniers." *Jacobin,* January 21, 2019. https://jacobinmag.com/2019/01/climate-change-2020-democrats-green-new-deal.

"Marshall Saunders." Climate Lobby. https://citizensclimatelobby.org/about-ccl/citizens-climate-lobbys-founder.

Master, Jeff. "A Look Back at the Horrific 2020 Atlantic Hurricane Season." Yale Climate Connections, December 1. 2020. https://yaleclimateconnections.org/2020/12/a-look-back-at-the-horrific-2020-atlantic-hurricane-center.

Maxwell, Angie. "How Southern Racism Found a Home in the Tea Party." *Vox,* July 7, 2016. https://www.vox.com/2016/7/7/12118872/southern-racism-tea-party-trump.

Maxwell, Angie. "What We Get Wrong about the Southern Strategy." *Washington Post,* July 25, 2019. https://www.washingtonpost.com/outlook/2019/07/26/what-we-get-wrong-about-southern-strategy.

Mayhew Bergman, Megan. "They Chose Us Because We Were Rural and Poor: When Environmental Racism and Climate Change Collide." *Guardian,* March 8, 2019. https://www.theguardian.com/environment/2019/mar/08/climate-changed-racism-environment-south.

Mayhew Bergman, Megan, "Why People in the US South Stay Put in the Face of

Climate Change." *Guardian,* January 24, 2019. https://www.theguardian.com/environment/2019/jan/24/us-south-climate-change.

McCright, Aaron M., Chenyang Xiao, and Riley E. Dunlap. "Political Polarization on Support for Government Spending on Environmental Protection in the USA, 1974–2012." *Social Science Research* 48, November 2014. https://www.sciencedirect.com/science/article/abs/pii/S0049089X1400132X?via%3Dihub.

McDonald, Jared, Bo MacInnis, and Jon A. Krosnick. "Climate Insights 2020: Opinion in the States." Resources for the Future. October 26, 2020. https://www.rff.org/publications/reports/climateinsights2020-opinion-in-the-states/.

McDonald, Tim. "The Man Who Would Be the First Climate Change Refugee." *BBC News,* November 5, 2015. https://www.bbc.com/news/world-asia-34674374.

McGurty, Eileen Maura. "From NIMBY to Civil Rights: The Origins of the Environmental Justice Movement." *Environmental History* 2, no. 3 (1997): 301–323. https://www.journals.uchicago.edu/doi/10.2307/3985352.

Meyer, Robinson. "The American South Will Bear the Worst of Climate Change's Costs." *Atlantic,* June 29, 2017. https://www.theatlantic.com/science/archive/2017/06/global-warming-american-south/532200/.

"Miami Residents Vote for $400 Million Miami Forever Bond." Associated Press, November 7, 2017. https://apnews.com/article/a05e8bbf60b34eb88648bb74234e6d38.

Middleton, Jennifer. "Battle of Carver Terrace." Pressreader. *Texarkana Gazette,* October 14, 2015. https://www.pressreader.com/usa/texarkana-gazette/20161007/281505045724979.

Miller, Zell. "How Democrats Lost the South." *Washington Times,* November 3, 2003. https://www.washingtontimes.com/news/2003/nov/3/20031103-123326-5341r.

Moline, Michael, "DeSantis Wants to Ban State Pension Investment Based on Political Factors." *Florida Phoenix,* July 27, 2022. https://floridaphoenix.com/2022/07/27/desantis-wants-to-ban-state-pension-investments-based-on-political-factors.

Mooney, Chris, Brady Dennis, Kevin Crow, and John Muyskens. "The Drowning South: Where Seas Are Rising at Alarming Speed." *Washington Post,* April 29, 2024. https://www.washingtonpost.com/climate-environment/interactive/2024/southern-us-sea-level-rise-risk-cities/.

Moser, Bob. *Blue Dixie: Awakening the South's Democratic Majority.* New York: Times Books/Henry Hold and Company, 2013.

Moser, Bob. "The Fearless Rise of the Black Southern Progressive." *New Republic,* October 19, 2018. https://newrepublic.com/article/151619/fearless-rise-Black-southern-progressive.

Mullin, Rick, and Cheryl Hogue. "The Rise of Environmental Justice." *C&EN, Chemi-*

cal and Engineering News 98, no. 32, August 24, 2020. https://cen.acs.org/environment/pollution/rise-environmental-justice/98/i32.

Mumgaard, Bob. "CFS Will Build Its First ARC Fusion Power Plant in Virginia" (blog). Commonwealth Fusion Systems. December 17, 2024. https://blog.cfs.energy/cfs-will-build-its-first-arc-fusion-power-plant-in-virginia.

Murawski, John. "Georgetown University's Road to Slavery Reparations Was Paved with Good Intentions, Leading to a Can of Worms." RealClear Investigations, Georgetown University. July 13, 2022. https://reparationscomm.org/reparations-news/georgetown-universitys-road-to-slavery-reparations-was-paved-with-good-intentions.

"The National Advisory Commission on Civil Disorders, Report, The United States Department of Justice, Office of Justice Programs, 1967." U.S. Department of Justice. National Advisory Commission on Civil Disorders. https://belonging.berkeley.edu/sites/default/files/kerner_commission_full_report.pdf?file=1&force=1.

Neidl, Phoebe. "How to Talk to People Who Doubt Climate Change." *Rolling Stone*, March 3, 2020. https://www.rollingstone.com/politics/politics-features/katharine-hayhoe-evangelical-christian-climate-scientist-953086.

Nesbit, Jeff. *This is the Way the World Ends: How Droughts and Die-Offs, Heat Waves and Hurricanes Are Converging on America.* New York: Thomas Dune Books, St. Martin's Press, 2018.

Nicoletti, Angela. "Mangroves Need Everglades Restoration Now More Than Ever." *FIU News*, Florida International University, June 22, 2022. https://news.fiu.edu/2022/mangroves-need-everglades-restoration-now-more-than-ever.

North American Moving Services. 2021 Migration Report. https://avlnavlblob.blob.core.windows.net/northamerican-com/docs/default-source/default-document-library/2021-migration-map.pdf.

"Obama Election Spurs Race Threats, Crimes." Staff written. NBC News, November 15, 2008. https://www.nbcnews.com/id/wbna27738018.

"Obama Polled Low with Southern Whites." *USA Today*, November 19, 2012. https://www.usatoday.com/story/theoval/2012/11/19/obama-southern-white-vote/1714291.

"On Mississippi's Gulf Coast, a Local Advocate Works to Protect Underserved Communities and Wetlands," Pew Charitable Trust, June 15, 2023. https://www.pew.org/en/research-and-analysis/articles/2023/06/15/on-mississippis-gulf-coast-a-local-advocate-works-to-protect-underserved-communities-and-wetlands.

Onwenu, Justin. "Trump Does Not Know What Environmental Injustice Looks Like." *The Hill*, November 19, 2012. https://thehill.com/opinion/energy-environment/523235-trump-does-not-know-what-environmental-injustice-looks-like.

Oreskes, Naomi, and David Gelles, "The Myth of Free Markets." Climate One Podcast, May 26, 2023. https://www.climateone.org/audio/naomi-oreskes-david-gelles-and-myth-free-markets.

Osborne, James, "Perry Says Texans Willing to Suffer Blackouts to Keep Feds out of Power Market." *Houston Chronicle,* February 18, 2021. https://www.houstonchronicle.com/business/energy/article/Perry-says-Texans-wiling-to-suffer-Blackouts-to-15956705.php.

"Our History," Virginia Community Rights Network. https://vacommunityrights.org/about/background.

"Our Mission," National Black Environmental Justice Network. https://www.nbejn.com/about-platform.

Owen, J. R., D. Kemp, A. M. Lechner, et al. "Energy Transition Minerals and Their Intersection with Land-Connected Peoples." Nature Sustainability6, December 1, 2022, 203–211. https://doi.org/10.1038/s41893-022-00994-6.

Parajon, Eric. "The Effect of Racial Resentment and Out-Group Cues on Support for Climate Policy." https://ericparajon.com/files/papers/The_Effect_of_Racial_Resentment_and_Out_Group_Cues_on_Support_for_Climate_Policy.pdf

"People of Color Spark the Environmental Justice Movement." *May 19th Project.* Public Broadcasting System. May 19, 2022. https://www.pbssocal.org/news-community/may-19th-project/people-of-color-spark-the-environmental-justice-movement.

Perls, Hannah. "EPA Undermines Its Own Environmental Justice Programs." Environmental & Energy Law Program. Harvard University, November 12, 2020. https://eelp.law.harvard.edu/epa-undermines-its-own-environmental-justice-programs.

Pezzullo, Phaedra C., and Ronald Sandler, eds. "Introduction. Revisiting the Environmental Justice Challenge to Environmentalism." In *Environmental Justice and Environmentalism: The Social Justice Challenge to the Environmental Movement,* edited by Ronald Sandler and Phaedra C. Pezzullo. Cambridge, MA: MIT Press, 2007. https://direct.mit.edu/books/book/4413/Environmental-Justice-and-Environmentalism.

Phillips, Kevin. *American Theocracy: The Peril and Politics of Radical Religion, Oil and Borrowed Money in the 21st Century.* New York: Viking, 2006.

Phillips, Kevin. "How the GOP Became God's Own Party." *Washington Post,* April 2, 2006. https://www.washingtonpost.com/wp-dyn/content/article/2006/04/01/AR2006040100004.html.

Pierre-Louis, Kendra, "If You Live in the South, Climate Change Could Kill Your

Economy." *Popular Science*, June 29, 2017. https://www.popsci.com/climate-change-economy-county.

Pittman, Craig. "FL Sacrificed a River to get Paper Mill Jobs. The Deal Really Stunk." *Florida Phoenix*, January 7, 2021. https://floridaphoenix.com/2021/01/07/fl-sacrificed-a-river-to-get-paper-mill-jobs-the-deal-really-stunk.

Plait, Phil. "Climate Denier Ted Cruz to Oversee NASA. Houston, We Have a Problem." *GRIST*. June 13, 2015. https://grist.org/politics/climate-denier-ted-cruz-to-oversee-nasa-houston-we-have-a-problem.

"Predictions Put Some of South Florida Under Water by 2025." CBS Miami, October 14, 2015. cbsnews.com/miami/news/predictions-put-south-florida-under-water-by-2025.

"Priority Program4: Use Green Infrastructure to Protect Environmentally Sensitive Areas and Integrate Nature into the City." Austin Texas Government. https://aura-atx.org/imagine-austin-priority-program-4-use-green-infrastructure-to-protect-environmentally-sensitive-areas-and-integrate-nature-into-the-city.

Proceedings: The First National People of Color Environmental Leadership Summit. https://www.cec.org/files/first-national-people-of-color-environmental-leadership-summit.pdf.

"Procter and Gamble's Buckeye Paper Mill Pollutes the Fenholloway River in Taylor Co., Florida, USA." Atlas of Environmental Justice. Institute of Environmental Science and Technology (ICTA), Universitat Autonoma de Barcelona, April 6, 2021. https://ejatlas.org/conflict/procter-and-gambles-buckeye-paper-mill-pollutes-the-fenholloway-river-in-taylor-county-florida.

"Public Good or Private Wealth?" Oxfam Briefing Paper, January 2019. https://oxfamilibrary.openrepository.com/bitstream/handle/10546/620599/bp-public-good-or-private-wealth-210119-en.pdf, p 12.

Railton, Stephen, with the University of Virginia Library. *Faulkner at Virginia*. May 7, 1958. Department of Psychiatry, tape 1, T-144a. University of Virginia. https://faulkner.lib.virginia.edu/display/wfaudio28_1.html.

Ramos, Carlos Nogueras. "Texas Energy Company Wins First-of-its-kind Permit to Suck Carbon Out of Air, Store Underground." *Texas Tribune*, April 8, 2025. https://www.texastribune.org/2025/04/08/texas-direct-carbon-capture-oxy.

"The Rate of Change since the Mid-20th Century Is Unprecedented over Millennia," NASA. https://science.nasa.gov/climate-change/evidence/.

Ray, Rashawn, and Andre M. Perry. "Why We Need Reparations for Black Americans." Brookings Institution, April 15, 2020. https://www.brookings.edu/policy2020/bigideas/why-we-need-reparations-for-Black-americans.

"Report of the Permanent 2022 Platform & Resolutions Committee." Texas State Republican Party. https://www.courthousenews.com/wp-content/uploads/2022/06/resolution-passed-by-a-texas-gop-party-committee-at-its-2022-convention.pdf.

Renkl, Margaret. *Graceland at Last: Notes on Hope and Heartache from the American South.* Minneapolis: Milkweed Editions, 2021.

Renkl, Margaret. "Lies, Damn Lies, and Georgia." *New York Times,* January 11, 2021. https://www.nytimes.com/2021/01/11/opinion/georgia-senate-democrats-south.html.

"Republican Platform 2016." The Republican Party. https://www.presidency.ucsb.edu/sites/default/files/books/presidential-documents-archive-guidebook/national-political-party-platforms-of-parties-receiving-electoral-votes-1840-2016/117718.pdf.

"Restoring the Quality of our Environment, Report of the Environmental Pollution Panel President's Science Advisory Committee." The White House, November 1965. https://nsarchive.gwu.edu/document/31937-document-2-white-house-report-restoring-quality-our-environment-report-environmental.

Ricketts, Sam, Rita Cliffton, Lola Oduyeru, "States are Laying a Road Map for Climate Leadership." Center for American Progress, April 20, 2020. https://www.americanprogress.org/article/states-laying-road-map-climate-leadership/.

Rifkin, Jeremy. *The Green New Deal: Why the Fossil Fuel Civilization Will Collapse by 2028, and the Bold Economic Plan to Save Life on Earth.* New York: St. Martin's Press, 2019.

Ripple, William J., Christopher Wolf, Thomas M. Newsome, Phoebe Barnard, and William R. Moomaw. "World Scientists' Warning of a Climate Emergency." *BioScience* 70, no. 1 (2020): 8–12. https://doi.org/10.1093/biosci/biz088.

Ritchie, Josh. "The Climate Doomsayers." *Politico 50, Our Guide to the Thinkers, Doers and Visionaries Transforming American Politics. Politico.* 2021. https://www.politico.com/magazine/politico50/2016/philip-stoddard-harold-wanless.

Ro, Alexander. "Why is Trump so Popular? Visiting the South Gave One Writer some Clues." *Guardian,* March 24, 2016. https://www.theguardian.com/books/2016/mar/24/donald-trump-popularity-deep-south-paul-theroux-book.

Roberts, David. "The GOP is the World's Only Major Climate Denialist Party. But Why?" *Vox,* December 2, 2015. https://www.vox.com/2015/12/2/9836566/Republican-climate-denial-why.

Roberts, Diane. "Florida is Turning Its Back on the New South, Embracing Its Dixiefied Past." *Florida Phoenix,* November 21, 2022. https://floridaphoenix.com/2022/11/21/florida-is-turning-its-back-on-the-new-south-embracing-its-dixie-fied-past/.

Rogin, Ali. "How the Mineral Mining Boom Endangers Indigenous Communities." PBS NewsHour, August 3, 2024. https://www.pbs.org/video/indigenous-mining-1722710980.

Roth, Mike. "Coming to Ballot in 2022? Protect Rights of Nature Through Florida Constitution." *TCPalm, Treasure Coast Newspapers*, July 23, 2021. https://www.tcpalm.com/story/opinion/contributors/2021/07/23/protect-rights-nature-waterways-florida-constitution-opinion/7890160002.

Rothstein, Richard. *The Color of Law*. New York: Liveright, 2017.

Samuels, Brett. "Biden Rips 'Beyond Ridiculous' Hurricane Claims from Greene." *The Hill*, October 9, 2024. https://thehill.com/homenews/administration/4924246-biden-rips-beyond-ridiculous-hurricane-claims-from-greene.

Sasso, Michael, and Alexandre Tanzi. "A $100 Bill Wealth Migration Tilts US Economy's Center of Gravity South." *Bloomberg News*, June 29, 2023. https://www.bloomberg.com/news/features/2023-06-29/millions-move-to-the-south-as-us-economy-favors-its-wealth-job-opportunities.

Schiermeier, Quirin. "Climate Science Is Supporting Lawsuits that Could Help Save the World." *Nature*, September 2021. https://www.nature.com/articles/d41586-021-02424-7.

Schleifstein, Mark. "Climate Change Poses Existential Threat to South Louisiana: U.N. Report." *Times-Picayune*, March 8, 2022. https://www.nola.com/news/environment/article_47392464-9e6a-11ec-84c6-1b558ec5b556.html.

Schleifstein, Mark. "Governor Edwards' Climate Task Force Unanimously Approves State Greenhouse Gas Reduction Plan." NOLA.com, January 31, 2022. https://www.nola.com/news/environment/gov-edwards-climate-task-force-unanimously-approves-state-greenhouse-gas-reduction-plan/article_ecb569ee-82da-11ec-b5b9-532f3fcfc196.html.

Schmitt, Mark. "The Myth of the Powell Memo." *Washington Monthly*, August 29, 2016. https://washingtonmonthly.com/magazine/septemberoctober-2016/the-myth-of-the-powell-memo.

"Secretary-General's Press Conference on Climate by Secretary-General Antonio Guterres at United Nations Headquarters." Press release. United Nations, July 27, 2023. https://www.un.org/sg/en/content/sg/press-events/2023-07-27/secretary-generals-press-conference-climate.

Sellers, Christopher. "Today's Anti-Environmentalism in Historical Perspective." H-EnviroHealth, Humanities and Social Sciences. Department of History, Michigan State University. https://networks.h-net.org/node/114427/discussions/180348/today%E2%80%99s-anti-environmentalism-historical-perspective.

The Sentencing Project. "The Color of Justice: Racial and Ethnic Disparity in State Prisons." Washington, DC. October 13, 2021. https://www.sentencingproject.org/reports/the-color-of-justice-racial-and-ethnic-disparity-in-state-prisons-the-sentencing-project.

Shabecoff, Philip. "Reagan and Environment: To Many, a Stalemate." *New York Times*, January 2, 1989. https://www.nytimes.com/1989/01/02/us/reagan-and-environment-to-many-a-stalemate.html.

Sharf, Samantha. "The Best States for Business 2019: North Carolina and Texas on Top." *Forbes.com*, December 19, 2019. https://www.forbes.com/sites/samanthasharf/2019/12/19/the-best-states-for-business-2019-north-carolina-and-texas-utah-on-top.

Shepherd, Katie. "Rick Perry Says Texans Would Accept Even Longer Power Outages 'to keep the federal government out of their business.'" *Washington Post*, February 18, 2021. https://www.washingtonpost.com/nation/2021/02/17/texas-abbott-wind-turbines-outages.

Silkenat, David. *Scars on the Land: An Environmental History of Slavery in the American South*. New York: Oxford University Press, 2022.

Silver, Timothy. "Three Worlds, Three Views: Culture and Environmental Change in the Colonial South." Nature Transformed, the Environment in American History. Appalachian State University. National Humanities Center. https://nationalhumanitiescenter.org/tserve/nattrans/ntuseland/essays/threeworlds.htm.

Smith, Adam. "2020 U.S. Billion-Dollar Weather and Climate Disaster in Historical Context." Climate.gov, January 8, 2021. https://www.climate.gov/disasters2020.

Smith, Danna. "Southern Communities for a Green New Deal." Press release. Dogwood Alliance, December 2, 2015. https://www.dogwoodalliance.org/2021/03/release-southern-communities-for-a-green-new-deal/.

Smith, Heather. "Climate Deniers Are More Likely to Be Racists. Why?" *Sierra*, June 18, 2018. https://www.sierraclub.org/sierra/climate-deniers-are-more-likely-be-racist-obama-trump-climate-change.

Smith, Jessie Christopher. "What the Record-Setting July Heat Means for the Rest of Oklahoma's Summer." *Oklahoman*, July 21, 2022. https://www.oklahoman.com/story/weather/2022/07/21/okc-forecast-record-breaking-high-temperatures-not-expected-august/65378589007.

Smith, Laura Ellyn. "The Spread of C-19 in the South Shows the Risks of Anti-Intellectualism." *Washington Post*, August 4, 2020. https://www.washingtonpost.com/outlook/2020/08/04/spread-covid-19-south-shows-risks-anti-intellectualism.

Solomon, Hsiang, Robert Kopp, Amir Jina, James Rising, Michael Delgado, Shashank Mohan, D. J. Rasmussen, Robert Muir-Wood, Paul Wilson, Michael Oppenheimer, Kate Larsen, and Trevor Houser. "Estimating Economic Damage from Climate Change in the United States." *Science* 356, no. 6345, 2017. https://www.science.org/doi/10.1126/science.aal4369.

"South Must Play a Key Role in Addressing Climate Change, Biodiversity Crises." Southern Environmental Law Center, April 15, 2021. https://www.southernenvironment.org/news/south-must-play-a-key-role-in-addressing-climate-biodiversity-crises.

"Southeastern United States Asian American Profiles (2022)." Asian American Center. University of North Carolina at Chapel Hill. https://aac.unc.edu/data-demography.

SouthWest Organizing Project letter to National Wildlife Federation and "Group of Ten" Environmental Groups, March 16, 1990. EJnet.org: Web Resources for Environmental Justice Activists. https://www.scribd.com/doc/26270824/SWOP-Letter-to-Group-of-10-Environmental-Organizations-1990.

"Southwestern States Have Better Solar Resources and Higher Solar PV Capacity Factors." U.S. Energy Information Administration, June 12, 2019. https://www.eia.gov/todayinenergy/detail.php?id=39832.

Sowell, Thomas, *The Quest for Cosmic Justice.* New York: The Free Press, Simon & Schuster, 1999.

"Study Finds Exposure to Air Pollution Higher for People of Color Regardless of Region or Income." Center for Air, Climate, and Energy Solutions. Environmental Protection Agency, September 20, 2021. https://www.epa.gov/sciencematters/study-finds-exposure-air-pollution-higher-people-color-regardless-region-or-income.

Summers, Brenda. "PCB Protest in Warren County 1982." WBTV News, Charlotte, NC, an affiliate of CBS News, September 16, 1982. https://www.youtube.com/watch?v=1iCxhoBYjgI.WBTV News.

Swartz, Kristi E. "Donors Have Stake in PSC Decisions." *Atlanta Journal-Constitution,* July 21, 2012. https://www.ajc.com/news/local/donors-have-stake-psc-decisions/ztFWBxrRTvJHg3gYjoYSMO.

Taheri, Mandy. "Millions of Americans Warned of 'Dangerous, Life-Threatening' Conditions." *Newsweek,* June 20, 2025.https://www.newsweek.com/millions-americans-warned-dangerous-heat-2088628.

Taylor, Chrishandra. "Rhetorical Appeals of Environmental Justice Communication." Technical Communication Capstone Course 6, 2015. https://cornerstone.lib.mnsu.edu/eng_tech_comm_capstone_course/6.

"Ten Years of Restoring and Building Trust 2012–2022. Land Buy-Back Program for Tribal Nations." US Department of the Interior, December 2023. https://www.doi.gov/sites/default/files/doi-lbb.pdf

"10th Anniversary of Hurricane Katrina." NOAA. Atlantic Oceanographic & Meteorological Laboratory. U.S. Department of Commerce, August 23, 2015. https://www.aoml.noaa.gov/10th-anniversary-of-hurricane-katrina.

Tessum, Christopher, Joshua S. Apte, Andrew L. Goodkind et al. "Inequity in Consumption of Goods and Services Adds to Racial-Ethnic Disparities in Air Pollution Exposure." *Proceedings of the National Academy of Sciences*, March 11, 2019. https://www.pnas.org/content/116/13/6001.

"Texas Coastal Resiliency Master Plan, March 2017." Texas General Land Office. https://www.glo.texas.gov/sites/default/files/coastal-grants/projects/files/Master-Plan.pdf.

The 3rd National Risk Assessment: Infrastructure on the Brink. First Street Foundation. October 2021. https://assets.firststreet.org/uploads/2021/09/The-3rd-National-Risk-Assessment-Infrastructure-on-the-Brink.pdf.

"13 Plead Guilty to Dumping PCB's in North Carolina." *New York Times*, June 15, 1979. https://www.nytimes.com/1979/06/05/archives/3-plead-guilty-to-dumping-pcbs-in-north-carolina.html.

Thompson, Andrea. "Florida's Record Warm 2015 an Outlier in U.S. East." *Climate Central*, December 21, 2015. https://www.climatecentral.org/news/florida-to-see-warmest-year-19860.

Tian, Ziyao, and Neil G. Ruiz. "Key Facts about Asian Americans Living in Poverty." Pew Research Center, March 27, 2024. https://www.pewresearch.org/short-reads/2024/03/27/key-facts-+about-asian-americans-living-in-poverty.

Tigre, Maria Antonia, and Margaret Barry. Climate Change in the Courts: A 2024 Retrospective. Sabin Center for Climate Change Law, January 2025. https://scholarship.law.columbia.edu/sabin_climate_change/240.

"Toxic Waste and Race in the United States. A National Report on the Racial and Socio-Economic Characteristics of Communities with Hazardous Waste Sites." Commission for Racial Justice. United Church of Christ, 1987. http://d3n8a8pro7vhmx.cloudfront.net/unitedchurchofchrist/legacy_url/13567/toxwrace87.pdf?1418439935.

Trask, Mililani. Video Personal Testimony, "The First National People of Color Leadership Summit." https://www.youtube.com/watch?v=tBT6oEWI5HI.

Traub, Alex. "Christopher Stone, Who Proposed Legal Rights for Trees Dies at

83." *New York Times,* May 28, 2021. https://www.nytimes.com/2021/05/28/us/christopher-stone-dead.html.

Tuckey, Mike. "Feds OK Mississippi's Katrina Grant Diversion." NBC News, January 25, 2008. https://www.nbcnews.com/id/wbna22805282.

Turner, Sylvester. Houston Climate Action Plan, City of Houston, 4. http://greenhoustontx.gov/climateactionplan/CAP-April2020.pdf.

"2008 Republican Party Platform." American Presidency Project. University of California, Santa Barbara. https://www.presidency.ucsb.edu/documents/2008-Republican-party-platform.

"2012 Republican Party Platform." American Presidency Project. University of California, Santa Barbara. https://www.presidency.ucsb.edu/documents/2012-Republican-party-platform.

"2019 Climate Action Summit." United Nations. https://www.un.org/en/climatechange/2019-climate-action-summit.

"2023 Defense Science Board Study on Climate Change and Global Security." Department of Defense, July 23, 2024, 13. https://dsb.cto.mil/wp-content/uploads/2024/08/DSB_SS2023_Climate_8-12-24_Cleared.pdf.

United States v. Robert Earl Ward, Jr. A/k/a "buck" Ward, 676 F.2d 94 (4th Cir. 1982). https://law.justia.com/cases/federal/appellate-courts.

University of Minnesota. "Biodiversity Survey Reveals That More Species Are Threatened with Extinction Than Previously Thought." *ScienceDaily,* July 18, 2022. www.sciencedaily.com/releases/2022/07/220718142320.htm.

"U.S. Climate Vulnerability Index Shows Where Action, Resources Are Most Urgently Needed." News release. Environmental Defense Fund, October 2, 2023. https://www.edf.org/cvi-news.

Vazquez, Nick. "Climate Treason March from Jacksonville to Tallahassee," August 21, 2021. https://www.facebook.com/photo/?fbid=353891242906957&set=a.103637527932331.

Virginia Center for Restorative Justice. "Our Programs." https://www.vcrj.org/Our-Programs.

Volokh, Eugen. "Foreign Law in American Courts." *Oklahoma Law Review* 66, no. 2 (2014): 220–221. https://digitalcommons.law.ou.edu/cgi/viewcontent.cgi?article=1056&context=olr.

Waldman, Scott, and Benjamin Hulac. "This Is When the GOP Turned Away from Climate Policy." ClimateWire. *E&E News,* December 5, 2018. https://www.eenews.net/articles/this-is-when-the-gop-turned-away-from-climate-policy/.

Walsh, Colleen. "Must We Allow Symbols of Racism on Public Land?" *Harvard Gazette*, June 19, 2020. https://news.harvard.edu/gazette/story/2020/06/historian-puts-the-push-to-remove-confederate-statues-in-context.

Washington, Harriet A. *A Terrible Thing to Waste: Environmental Racism and Its Assault on the American Mind.* New York: Little, Brown Spark, 2019.

Waters, Hannah. "The Flight of the Spoonbills Holds Lessons for a Changing Everglades—and World." *Audubon* magazine, Winter 2022. https://www.audubon.org/magazine/winter-2022/the-flight-spoonbills-holds-lessons-changing.

Watkins, Kim. "Houston's 'Father of Environmental Justice' on Where He Sees the Movement Headed." *Houston Public Media*, April 26, 2021. https://www.houstonpublicmedia.org/articles/news/energy-environment/2021/04/26/396685/what-is-environmental-justice-what-is-the-future.

We Choose Now: Climate Justice from the Southern Frontlines, a Conversation with Colete Pichon Battle. UC Santa Cruz Institute for Social Transformation. Video. https://transform.ucsc.edu/events/wechoosenow-climate-justice-from-the-southern-frontlines/.

West Virginia Climate Alliance. https://www.wvclimatealliance.org/about.

"What Are Coastal Wetlands?" Environmental Protection Agency, https://www.epa.gov/wetlands/about-coastal-wetlands

"What is Climate Resilience?" Union of Concerned Scientists, June 6, 2022. https://www.ucsusa.org/resources/what-climate-resilience.

"What's the Difference Between Climate Change Mitigation and Adaptation?" World Wildlife Fund. https://www.worldwildlife.org/stories/what-s-the-difference-between-climate-change-mitigation-and-adaptation.

"What's Wrong with the Argument 'The Climate Is Always Changing?'" *All Things Considered*, National Public Radio, December 12, 2018. https://www.npr.org/2018/12/12/676198899/climate-scientist-says-argument-the-climate-is-always-changing-is-wrong.

"What We Do—Foundation For Louisiana." https://www.foundationforlouisiana.org/what-we-do.

Whitehead, Frederika. "The First Climate Justice Summit: A Pie in the Face for the Global North." *Guardian*, April 16, 2014. https://www.theguardian.com/global-development-professionals-network/2014/apr/16/climate-change-justice-summit.

"White Nationalists, Other Republicans Brace for 'Total War.'" Southern Poverty Law Center, December 11, 2022. https://www.splcenter.org/hatewatch/2022/12/11/white-nationalists-other-Republicans-brace-total-war.

"Who Is APEN?." Asian Pacific Environmental Network. https://apen4ej.org/.
Williams, Angela. "'A Great Day to Not Be in Jackson,' Governor Says During Hattiesburg Event." WAPT, American Broadcasting Company. Jackson, MS affiliate, September 16, 2022. https://www.wapt.com/article/a-great-day-to-not-be-in-jackson-mississippi-governor-says-during-hattiesburg-event/41252344.
Wilson, Charles Reagan. "Religion and the US South." Southern Spaces. Emory Center for Digital Scholarship, March 16, 2004. https://southernspaces.org/2004/religion-and-us-south.
Yarina, Lizzie. "Your Sea Wall Won't Save You." *Places Journal*, March 2018. https://placesjournal.org/article/your-sea-wall-wont-save-you/.
"Yet Another Hurricane Wetter, Windier, and More Destructive Because of Climate Change." World Weather Attribution, October 11, 2024. https://www.worldweatherattribution.org/yet-another-hurricane-wetter-windier-and-more-destructive-because-of-climate-change.
Yates, Diana. "Changing the Density of Stomata to Improve Their Control of CO2 Influx and Moisture Loss. Not just CO2: Rising Temperatures Also Alter Photosynthesis in a Changing Climate." Plant Stomata Encyclopedia. Illinois News Bureau. University of Illinois, 2021. https://plantstomata.wordpress.com/tag/diana-yates.
Yuccas, Jamie. "Learn the Meaning of Environmental Racism," NickNews, video posted by Nickelodeon (@Nickelodeon), X, April 23, 2021, https://x.com/Nickelodeon/status/1385611519665246208.

Personal Interviews

Steve Adler, former mayor of Austin, Texas. Telephone interview. December 12, 2022.
Emilia Aguinaga, Climate Justice Program officer, Foundation for Louisiana. Telephone interview. June 6, 2025.
Aaron Ambroso, codirector and cofounder of Houston Climate Justice Museum. Telephone interview, April 14, 2023.
Zach Baumer, Climate Program manager, City of Austin, Texas. Telephone Interview. October 5, 2022.
Jan Booher, project director on an EPA Environmental Justice Grant awarded to the Green Sanctuary Committee of the Unitarian Universalist Fellowship of Boca Raton, Florida. Telephone interview. May 11, 2021.
Casi Calloway, chief resilience officer for the City of Mobile, Alabama. Telephone interview. May 25, 2023.

Alys Campaigne, climate initiative leader, Southern Environmental Law Center. Email exchange, February 7, 2025.

Judith Curry, professor emeritus and former chair of the School of Earth and Atmospheric Sciences at Georgia Institute of Technology. Telephone interview. December 19, 2022.

Andrea Dutton, a 2019 MacArthur Fellow and professor of geology in the Department of Geoscience at the University of Wisconsin–Madison. Telephone interview. 2017.

Alex Easdale, executive director of the Southeast Climate and Energy Network. Telephone interview. November 10, 2022.

Ryan Freed, climate action officer for Alexandria City Government. Telephone interview. November 14, 2023.

Daniel Gilford, climate scientist with Climate Central. Telephone interview. November 9, 2023.

Lisa Hall, media & community relations adviser at Babcock Ranch, Florida. Personal interview. April 9, 2024.

Michael Hansen, former executive director of Greater-Birmingham Alliance to Stop Pollution (GASP). Telephone interview. December 3, 2021.

Matt Hauer, assistant professor of sociology at Florida State University, climate migration specialist. Telephone interview. October 26, 2022.

John S. Jackson, political scientist and former chancellor of Southern Illinois University Carbondale. Telephone interview. May 30, 2025.

Stefan Lehne, Senior Fellow at Carnegie Europe. Email exchange. December 11, 2022.

Susan MacManus, University of South Florida Distinguished University Professor Emerita (political science) and political analyst. Telephone interview. November 11, 2021.

Margie Vicknair-Pray, conservation program coordinator, Sierra Club, Delta Chapter. Email exchange. October 24, 2022.

Codi Norred, executive director at Georgia Interfaith Power & Light. Telephone interview. October 25, 2021.

Ben Price, national organizing director of the Community Environmental Legal Defense Fund (CELDF). Telephone interview. December 3, 2021.

Scot Quaranda, communication director, Dogwood Alliance, Asheville, North Carolina. Telephone interview. October 20, 2022.

Virginia Sanders, National Sierra Club Environmental Justice employee and a member of the Lower Richmond, South Carolina NAACP. Telephone interview. November 4, 2023.

Ryan Smart, executive director at Florida Springs Council, Tallahassee. Telephone interview. October 1, 2021.

Kyle Spencer, chief resilience officer, City of Norfolk, Virginia. Telephone interview. May 1, 2023.

Philip Stoddard, former mayor of South Miami, White House Governance Coordinating Committee of the National Ocean Council, and biology professor at Florida International University. Telephone interview. December 11, 2022.

Jack Sweeney, former campaign assistant with the Louisiana Bucket Brigade. Telephone interview. March 24, 2023.

Chandra Taylor, senior attorney and leader of Southern Environmental Justice Initiative, Southern Environmental Law Center, Chapel Hill Office. Telephone interview. November 3, 2021.

L. Donelson Wright, former director of marine science at William & Mary's Virginia Institute of Marine Science. Telephone interview. February 2, 2023.

Gary Zarillo, professor of oceanography and ocean engineering, Florida Institute of Technology, Melbourne. Telephone interview. April 13, 2003.

INDEX

John M. Dunn has written more than 400 articles for periodicals such as *Europe, Overseas Life, Sierra, Florida Trend, Business Florida,* and the *St. Petersburg Times.* He has also authored 17 nonfiction young adult books, including *The Russian Revolution, The Relocation of the North American Indian, The Spread of Islam, The Civil Rights Movement,* and *José Martí: Cuba's Greatest Hero.* Several of his books have received national recognition. In February 2020, his book *Drying Up: The Fresh Water Crisis in Florida,* published by University Press of Florida, was selected as the winner of the Florida Historical Society's Stetson Kennedy Award for writing about Florida's natural environment. The book also won the Bronze Medal for Florida nonfiction in the 2019 Florida Book Awards. He lives with his wife in Ocala, Florida.